Towards Emerging Facets of Disaster– Emergency– and Crisis– Management

Towards Emerging Facets of Disaster–Emergency– and Crisis– Management

By:

Professor D. K. Sinha

Sir Rashbehary Ghose Professor of Applied Mathematics, (Retd.)
University of Calcutta
Formerly, Professor of Mathematics, Jadavpur University,
Formerly, Coordinator, Centre for Atmospheric Sciences,
University of Calcutta,
Past General President, Indian Science Congress Association.

Researchco Book Center

New Delhi - 110 005

First Published 2006

ISBN : 81-891402-9

Published by :

Researchco Book Center
25B/2, New Rohtak Road
(Near Liberty Cinema)
Karol Bagh
New Delhi-110005

Printed at :

Salasar Imaging Systems
New Delhi

ACKNOWLEDGEMENT

For a publication of this kind, one cannot but draw upon the vast torrent of literature on its theme. I hereby acknowledge sincerely what I could have as rich resources, predominantly intellectual from classical materials on relevant topics. I need to acknowledge what I could obtain from the valuable publications of NDO (Australia), UNESCO, UN, IDNDR secretariat, WCDR, publications on natural disasters, emergencies, crises etc.

CONTENTS

CONTENTS

PREFACE

It is a difficult proposition to trace back the beginnings of pursuits on disaster management. That man has battled with the aftermath of natural disasters also goes back to legendary days of the oft-cited Noah's Arc. Such accounts on the courage, tenacity along with sufferings of people affected, continue to be dismal, but not with that fatality as we find in the later half of the last century; contending with them seems to be in many ways diffusive. It is only few organisations and individuals,. as well who could assiduously accumulate information on disasters, and these are not visible in remarkable ways till we reached the seventies of the last century. It was the declaration by UN on International Decade on Natural Disaster Reduction (IDNDR) on the eve of the nineties that provided an enabling ethos to consolidate on available disaster knowledge for shareable purposes across the globe. Management of disasters could thus find a place of tip on the framework of the totality of work on natural disasters. Few words like emergencies, crises, contingencies etc. have irresistibly come to be allied with natural disaster management. Management exercises *per se* cannot but obviously join the fray. So whatever we may seek to designate, disaster–, emergency–, crisis–, contingency-management have all come into vogue ; in a way, disaster management is fast evolving with its varieties of characteristics and has come to acquire a niche in the annals of endeavours to align with the people amidst disasters. A natural disaster, if not always overwhelmingly devastating, like the recent tsunami, affords an opportunity to look at afresh, disaster management with its strengths and weaknesses. The present book is sought to be an attempt to look at the whole conspectus of disaster management having an evolutionary bearing on what has happened recently.

The conceptual framework of management sequences can hardly be dispensed with, when we talk about disaster management in terms of current parlance in management. The present treatise refrains from trivialising disaster management from the exercise of grafting on usual cliches. It does seek to look for conceptualising a natural disaster phenomenon, the recent tsunami-happenning being significantly the lurking motivation. The allied aspects, namely, crises, emergencies and contingencies are dealt within the purview of managerial frameworks. Risks and vulnerabilities in regard to natural disasters need to be taken on by any endeavour on managing disasters. Management of disasters often suffers from a subterfuge, presumbly because of bureau-politics and the reality gets blurred. This is all the more distressing if the forecasts and warning systems are reneged on. Environmental management has come to stay, but managing environmental degradation on account of cataclysms hasn't been so far clinched; the slow onset disasters like floods and droughts have a socio-political basis and so, keep on hanging with issues and dilemmas to the extent these can pay off and hence, managing a coupled character of disasters becomes an onerous and a challenging task. A few glimpses in these directions are sought to be delineated in this book. Many aspects for obvious reasons remain unsaid and are far from being delved into. On the other hand, neither a hindsight nor a foresight can explore or resurrect what have happened in terms of futurities, exhaustively. But the emerging developments, particularly those with an inbuilt structural innovation can, in no way, be precluded ; for example, knowledge vis a vis creativity have to have new vistas in the arena of disaster management. That's why a venture on envisioning disaster management so that it can move out of archaic ruts of applicable managerial skills and thereby bogged down to shelter and rescue of the people affected, but enthuse others to plunge themselves in the aftermath of disasters.

Environmental management agency has come up elsewhere ; diversification, devolution and dealings are often defined sectorally in terms of what the participation of the community is to be structured and appositely translated into realities. Global trends are in no way stoppable and one should draw upon them without being victims of inhibitions of communities at risk The Hyogo Declaration at Kobe in the early part of this decade followed by South Asian Summit in May 2005, have well focussed the areas of scenarios. The present book is to be treated as a facilitator to developing the culture and ethos, which is sorely missing around the orbits of natural disaster management.

The book assumes the details and technicalities of natural disasters and some revealingly significant areas and themes. For this, the other recent book "Towards Basics of Natural Disaster Reduction" authored by me may be taken as complementary or supplementary to the present book.

Any tract on disaster management, on the face of it, is mostly a purveyor of success stories which are, of course, noteworthy. Complexities of emergencies galore, the massive scale of the task for every event legitimately brings out the inadequacy of erstwhile approaches. Disaster policy so far, be it in the developed or developing world, often proves to be neither necessary nor sufficient. The recent tsunami events have proved in unmistakably clear terms, the terrible woes of the affected, to combat which endeavours have been far from being successful in respect of natural disaster reduction. There are somewhere gaps or obsessions with some sort of orthodoxy. While knowledge economy has come to stay, knowledge gap needs to be encountered by multifaceted strategies shaped by a comprehensive development framework, as harped on by Hyogo Declaration. Why shouldn't there be a campaign for a major change in thinking, a shift towards improving the state of affairs ? As two former Presidents of USA have remarked very recently, it is still irksome to find that the tsunami support has virtually gone by default. Failed states should, in no way, be treated to as impediments to development.

In a country like ours, the formation of a Disaster Management Agency is well on the offing as an upshot of legislative processes. Central response and assistance will hopefully go a long way in establishing a process and structure for the systemic, coordinated and effective delivery of central assistance to address the consequences of any major natural disaster or emergency, ostensibly declared under the proposed Act. The proposed agency will certainly provide for inter-agency and inter-governmental emergency preparedness and planning, training, exercising, coordination and information exchange. The scope for the agency is, undoubtedly, enormous and so, is required a studied identification and analysis. The organisational framework has to be necessarily looked into. There cannot but be its continuing actions and conceptualisation of operations. The proposed agency will certainly be a new venture on finding a balanced role for the states and the centre which together must evince, an understanding of the limitations of both and also of the corporate sectors. Not any shred of conditionality from any organisation will work, as there are likely to be legitimate disagreements on key issues. Dilemmas, debates and furore can well ensue on matters of agenda, if not on ideology. But what is most essential is that the country concerned must be put in the driver's seat. In the ultimate analysis, if the duration of the disaster be of enormous change, tumult, an opportunity, there should be a commitment for the developing world, in particular, to leave an impressive legacy for their successors to take up.

I must convey first my heartfelt thanks to the publisher Researchco Book Center of New Delhi for prevailing on me to have two books basically on natural disasters, this being the second one authored by me. I must also say that I have seldom come across a friendly publisher under the leadership of Sri N.K. Jain, Sri Anil Jain and Sri Arvind Jain. Of course, the proposal was initiated by Sri Prasanta Saha of World Book of Kolkata and Sri Aditya Biswas, the Librarian of All India Institute of Public Health & Hygiene, Kolkata. I know it for certain that this is not enough for expressing my gratitude to all of them.

Since I had to undertake the pressing task and also, doubtless, a challenging one, I had to turn to Sri Basudeb Chatterjee who helped me enormously as amanuensis for this purpose and sincere thanks are due to him. The technology had to be resorted to in keeping with the needs of the day as a provider to the printer and for the technical advice, I had to seek it from Sri Devaprasanna Sinha and Dr. Devadatta Sinha and so, thanks to them. Finally, my thanks are in some deeper ways, due to the members of my family who did put up with the discomforts that often accompany this onerous work within a limited period. I look forward to inputs and feedback on what I penned down through the book.

Kolkata, Buddha Purnima,
May 23, 2005

D.K. Sinha

1

PROLOGUE : WHITHER CONTOURS OF NATURAL DISASTER MANAGEMENT?

Introduction

No aspect of management has so far been talked about with naivety as that dealing with natural disasters. The reasons are not, perhaps, far too seek. It looks that even the consideration of such types of management during the decade of nineties, labeled as International Decade for Natural Disaster Reduction (IDNDR), has not been able to endow such pursuits with a truly management flavour. The simplest of such disabilities is that the policy makers and the political community may keep on exacting apolitical, if not often political, mileage of anything to do with natural disasters, particularly-the case of relief. The message of IDNDR is that the accent should be more on pre-disaster phase rather than post disaster phase. The confusion on this score, which still persists, has allowed many sectors to join the fray. It has come to such a pass that even from the point of view of management, *per se,* managers read more of their expertise than issues and problems of disasters in their entirety. People from bureaucracy or what is often called, those of bureau-politics tend to look at it as management of crisis ; there has been no dearth to treat of disasters as those with emergency characters which have essentially managerial overtones. Those exists a plethora of literature on natural disasters, vide, Sinha [1], [2], Drabek [4], Sinha & Mukherjee [3], Visva-Bharati [5], Davis [6], Fischer III [7], along with those by UN and their affiliates, for example, PAHO, WHO [8]. Any kind of study or activity about disasters cannot but be interdiscplinary and hence, managerial flair on pursuits of (natural) disasters has to draw upon science, technology, behavioral studies, human psyche, health and, perhaps, what not. What is solely missing is the identification of contours of management in regard to natural disasters and this is what the present chapter seeks to do. Essentially, it presents a kind of critique of what have gone by to grapple with and to contend with, natural disasters, to analyse and to study, what these are all about in the singularly three distinct phases. Having done this, the presentation is wound up by indicating new directions of natural disaster management that fit in with its emerging facets. That natural disaster management endeavours ought to acquire paradigms in the context of their overarching characteristics not only from a conceptual and structurisational standpoint but also from the impending incursions of contemporary technologies so as to cater to enabling services at the ground level, has also to be reckoned.

Some Basic Aspects of Natural Disaster Management

Any preliminary statement on disaster management cannot but be simplistic. There is every likelihood of two terms "disaster" and "hazard" being taken synonymously. In order to obviate this possible confusion, the term "disaster" has to be taken to refer to the natural event ; for example, an earthquake along with its adverse effects, e.g. the loss of life and/or destruction of properties, while the "hazard" just refers to the natural event. Another term used invariably with the above two terms is "vulnerability", which refers essentially to the susceptibility of the system, e.g. water supply or any infrastructural aspect related to the effects of the hazard. If we are to associate management of these entities even with some kind of naivety, we must reckon what is called "disaster management". These are disaster preparedness, disaster mitigation and disaster response. Obviously, in a management sequence of a disaster, the impact of a disaster followed by rehabilitation and reconstruction, besides the above elements, become well-structured. IDNDR has stressed, in unambiguous terms, elements of natural disaster management to be taken up at different levels with multi-sectoral responsibilities. It must be mentioned that health disaster management can hardly be ruled out for such scenarios, without being delinked from necessities of environmentally sensitive aspects of health. The usual rudiments such as planning, layout, temporary settlement camp etc. come within the purview of management of disasters ; communication and transport need to be vigorously managed making use of all forms of latest developments of communication technology e.g. which can do without radio communication in regard to High Frequency (HF) Single - Side (SS) radio, Very High Frequency (VHP) hand held radio communication, amateur radio operators, terrestrial and satellite communication with all ramifications, particularly for the latter, electronic mail services. The human element lies perennially been with humanitarian relief services and management on such scores keeps on evolving. In all these managerial sectors, there are many which may become amenable to management within the run-of-the-mill strands of management, if taken ritualistically. These forms of management obviously acquire new dimensions, only when the elements of crisis, emergency or contingencies are taken into account.

Crisis management, whatever be its variants, continues to labour under, particularly in countries like ours, scepticism and misdirected applications.

That crisis management should reckon behavioral responses of dominantly affected group(s) has not been looked into seriously neither by the bureaucracy nor by the policy makers. Since behavioral responses are inextricably linked up with cognitive processes, crisis management has often gone by default, not merely conceptually but also in practice. Even the significance of communication technology or in terms of the current usage, information technology has not been conceptually dealt with in matters of crisis management. Since behavioral studies keep on evolving, management with a behavioral content, has to be continually updated. Emergency management has strong overlaps with crisis management. Even if this may have, on the face of it, simplistic overtones, it acquires complexity when issues of mitigation and risks are brought to focus. The sociology of disasters and hence, managing disasters taking into account the responses in their true totality, is fast assuming importance. Organisational management is another facet which has not been deeply delved into, in the light of recent developments on network technologies. The stark realities affecting management efforts are, in many ways, dominantly sociological. The social disruptions brought about by occurrences of disasters may well turn out to be social burdens and hence, these need to be treated with the methodology of societal management but with a difference. The mythical obsession about natural disasters shouldn't be brushed aside ; the whole gamut of disaster management ought to recognise this continuing influence of myths confronting the realities. In fact, the disaster-management community is inescapably coming up as a distinctive fallout, which Afro-Asia subcontinents are endeavouring with tremendous strides.

Wrap Up

Having presented briefly some glimpses on how to go about managing the system of natural disasters, it would be in the fitness of things to have a sort of critique on them so that the corresponding emerging or potentially emerging, traits of management can be focussed. What is missing in the extant approach to management of this kind is not being able to regard it somewhat akin to managing complexity with technical nuances. This has become all the more compelling because of the fast emerging "science of complexity" as a discipline. The segmented approach which does not treat the system of disasters as a whole, has not paid dividends ; the commonalities among each of the natural disaster-features ought to have wrought, by now, a change in approach. This distinctiveness of approach for a single disaster can in no way enable policy makers to develop the apposite management in action, far from what the conventional strategic and operational management may possibly provide. Hence, the need of reckoning the overarching view of complexity in managing the system of natural disasters. For example, managing droughts in Asia and Africa shares necessarily some dominant aspects of the corresponding managing system for droughts in the Indian context. Isn't it again a complexity viewpoint in keeping with the tenets of

IDNDR ? What ought to be taken up now is to look at, for example, those salients and snags of managerial enterprises in the Asian sub-continent. Shouldn't indepth pursuits on Asian affairs have within its fold this approach to natural disaster management ?

Some efforts of the establishments, wherever they may be, seem to have the propensity to deal somewhat casually about emergency management, having close interfaces with natural disaster management ; these are often referred to as contingencies and/or crisis management in wider perspectives. That crisis in regard to natural disaster management has a distinctive reality, is often lost sight of ; the understanding of the related uncertainties is also not adequately realised. Crisis becomes invariably a characteristic of "urgency" which, in turn, is felt subjectively, by policy makers. The decision makers are often led to some sort of a predicament because of not being capable of coping with emergency situations. What is required is to look upon crisis as a process rather than an event, alienated from the processes. Crisis management thus needs to be considered in processual terms. We can, perhaps, thereby make the practice of crisis management free from the ritualistic practices that we often see around. Shouldn't we make professionalism immune from the psychological and physiological pressures of crisis, without defying the intrinsic rationality involved ? Crisis management, contrary to what is being followed now, should look for modes of intervention going well beyond the just requisites of the situation. Information system, doubtless, plays a vital role for disseminating information of different dimensions of crisis management. Such considerations of crisis management lead to trans-nationalisation of contingencies and crises which, are nothing but sequels to the salients of IDNDR. Risk, crisis and disaster management have to have, if not yet by now, a composite entity.

Globalising natural disaster management becomes, therefore, a natural corollary of the message of IDNDR. How to go about integrating IT into global crisis management and then, evolving a natural disaster policy ? In particular, with the appropriate design and implementation of IT, natural disaster risk can be substantially reduced whether they occur at Bhuj or Teheran (earthquakes). We often and rightly so, speak of empowering communities prone to natural disasters. Can we design for community-information systems ? There are significant advances in IT so as to allow communities to live with seismicity, floods or droughts, without devastating losses of life and property. Can we now think of on a wide scale ITeS to fill in the necessities so as to bring about cadres or nucleus of cadres of enablers for natural disaster services ? Shouldn't there be an ilk of it in enterprises on management ?

Finally, a few words about organisations that fall within the ambit of the system to be managed. The governmental, the semi-governmental, regional or global, forms do participate regularly in coping with natural disaster systems ; NGOs and the emerging civil societies in different parts of the world also participate, often excelling those of the established work-force. Natural disaster management has, doubtless, a multiplicity of interests and groups and hence, there is a need for evolving, again, in view of complexity, a new sector of management. This form of management may be resented by NGOs but some sort of linkage has become an urgent necessity. The contexts of development and relief need to rope in organisations including NGOs in the roles of implementers, catalysts and partners and hence, the management has to work ceaselessly for developing necessary genre and ethos of natural disaster management without being indifferent to human touch, ethics and ethnicity, too. Managing enterprises on natural disasters has, no doubt, acquired leading heights but yet to have paradigms with significantly interdisciplinary traits. Environmental management, because of its continuing interest in human elements, can hardly be delinked from the kind of management being dealt here. Shouldn't the pervasive ideas and activities on human rights prevail on distressed communities, to ask for appropriately legitimate rights, because of their *insitu* vulnerabilities ? Managing disasters can ill afford to be indifferent to such potentially live issues.

2

ON THREE C'S : CONCEPT, CAUSE AND CONSEQUENCE OF A NATURAL DISASTER–THE RECENT TSUNAMI AS AN EXEMPLAR

Introduction

It has become a truism to say that natural disasters have several commonalities, mutualities but diversities in their adverse impacts. We seem to be fascinated with the largessee of aid, relief and the like, in the aftermath of the disasters. One cannot deny the compelling nature of circumstances–a disaster puts the community with its onset. It is not that there are people who carry on, in concrete terms, strategising recovery, renovation and rehabilitation plus something more than of an abiding nature. There are pursuits, undertaken by few, on what are often called lessons on the disaster. Apart from their practical spin off's, there are select few who delve deeper into causes and consequences of such disasters. It often looks that the conceptualisation of the whole process regarding disasters, takes a backseat, and therefore, this becomes of no use in the upshots derivable from a disaster.

As the recent tsunami on December 26, 2004, is being held, as a matter of grave concern, it serves as an adverse calamity that can well motivate its analysis and for others as well, in limited ways for consideration of three c's namely, concept, cause and consequence ; indeed, there can have another set of three c's namely, crisis, cataclysm and chaos, the last two being often taken as extremes of the first one. Coming back to the purpose of this chapter, we shall venture to go deeper into the three c's of a tsunami, amenable to possible generalisation of other forms of a natural disaster.

Conceptual Genesis of a Tsunani

Although, a bit technical, the run-of-the mill idea of a tsunami, being floated, is a tidal wave, which hardly connotes anything purposeful. A tsunami has its defining moments only when there is a tumultous sea-bed caused by seismicity of enormous severity. There is thus a kind of coupling of fault movement causing earthquakes occurring in the seabottom and generating a large wave. In brief, there is somewhere a conception because of a get together of two phenomena resulting in a product. Therefore tsunami, as a concept, can in no way be bereft of interactions ; the concept of a tsunami can in no way labour under a type of absolutism or objectivity which the use of the word 'concept' mostly conveys. A tsunami, like what we have witnessed recently, has burgeoning characteristics from the moment it becomes deceptively visible in the open sea. Another defining characteristic of a tsunami is shown by its movement along a shore line through an withdrawal of the water, as though by an extra-ordinarily low tide.

Conceptually speaking, tsunami, seismic sea waves or earthquake tidal waves are often juxtaposed, although etymologically speaking, tsunami is really the Japanese word, *tsu-nah-me* , meaning a *harbour wave*. It should not be taken to mean that all sub-oceanic seismic shocks are accompanied by movements of the sea bottom and so, the tsunami ; in fact, statistics bears testimony otherwise. A discernible feature of a tsunami is that the first motion is invariably a precursor to large waves in areas from where inhabitants are capable of fleeing away on a sort of non -technological warnings ; one cannot, however, a say anything about the time between the recession of the first wave and the arrival of the others. Apparently there exists a kind of capability of people who seem to be aliterately wise in regard to tsunamis or other types of coastal vulnerabilities.

Casuative Components for a Tsunami

It is not at all a different proposition to trace the causes of a tsunami. That it stems from seismicity is well known but extensive landslides or volcanic eruptions can be root causes of tsunamis, is not that well talked about. The popular term about it as a tidal wave is incorrect. There are many stories rather accounts of life threatening tsunamis occurring in the Atlantic ocean since 1918 but they pose serious threats in the Pacific. Even if, the tectonic configuration of a region, for example , the Caribbean basin, makes it susceptible to seismic activity, the earthquakes there rarely become tsunamigenic. It is told that the earliest recorded tsunami in Latin America occurred in 1691, inundating 1500 km of Chilean coastline. One can cite names of various places in Latin American countries, where devastating tsunamis occurred.

Let us look into the science of tsunami generation. Assuming a tsunami to have occurred as an event with a normal dip-steep fault movement with shallow hypocentre we know that the waves after being visible for a very short length in mid-ocean, they travel at incredible speeds and with great distances, they will be the wave crests.

A tsunami's speed decreases as the waves approach shallow water and the energy converts itself into wave height ; often reaching as high as 25 mts ; of course, the interval of time between successive waves remains unchanged, normally between 20 to 40 minutes. As tsunamis approach the coastline, there is withdrawal of the sea, often to levels lower than low tide, then abruptly rise as a giant wave. The configuration of the local shore line on the sea-bottom influences largely this sort amplification of the wave. A study of such effects is often made and it is found, on some occasions, that because of the source of the wave, the runup can go up to an height substantially higher than the crest of the wave at the shore line. (Fig. 2.1)

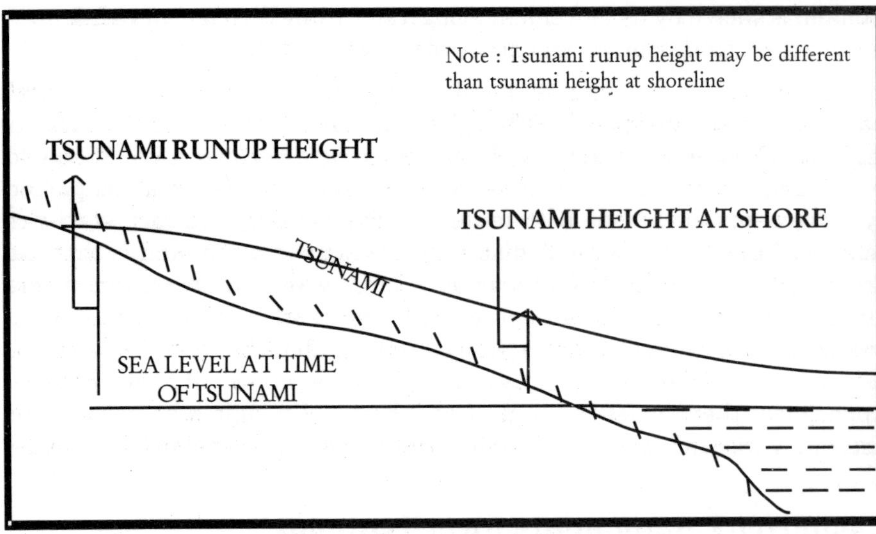

Fig. 2.1 : Tsunami Runup Height & Height
(Courtesy: Organization of American States, USAID)

In quantitative terms, if we assume an approximate water depth of about 365 mts. around the epicentre, then this will generate a tsunami wave with a velocity of 225 km per hour at the point of generation. This means that for shore areas about 10 kms from the epicentre, there would only be a lead time of 3 to 5 minutes from the time of earthquake occurrence to the strike of the tsunami, supposing the wave velocity to decrease as water depth changes near the beach. When the wave travels over shallow waters thereby conserving the energy of the wave, it strikes as a normal for refraction to occur thereby resulting in a greater wave amplitide at the shallow coastlines. Of course, the coastal topography will certainly govern the amplitude changes. There is a general pattern, shown by observations, that the average ratio of initial rise to succeeding fall is about 1 : 3 ie. withdrawal seems to be most striking if the true front of the wave is relatively a small crest.

Tsunamis are found to have caused damages in locations which have either narrow outlets or have broad sloping topography, so that runup bed friction comes up which, in turn, causes amplification of the wave height. The bathymetry and local coastline morphology determine the magnitude of the devastation and height of the tsunami wave from place to place. Any account of tsunami conjures up what happened following the Alaska earthquake as shown by the propagation of the tidal wave on Monday, April 1st, 1946, [Fig. 2.2). The tsunami is traced by tide gauges and seismographs.

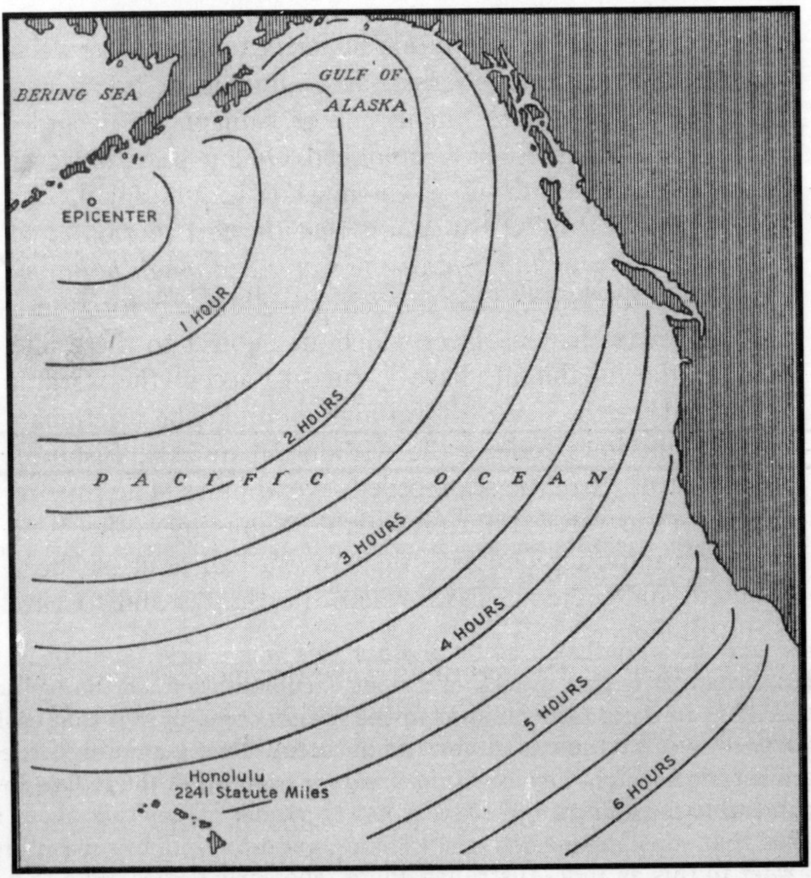

Fig. 2.2

The figure in the next page shows tsunami travel chart. (Fig. 2.3)

The recent Tsunami in the Indian Context

The recent tsunami occurred on 26th December, 2004, in the Indian ocean. There are many queries that remain still fresh in the minds of the people.

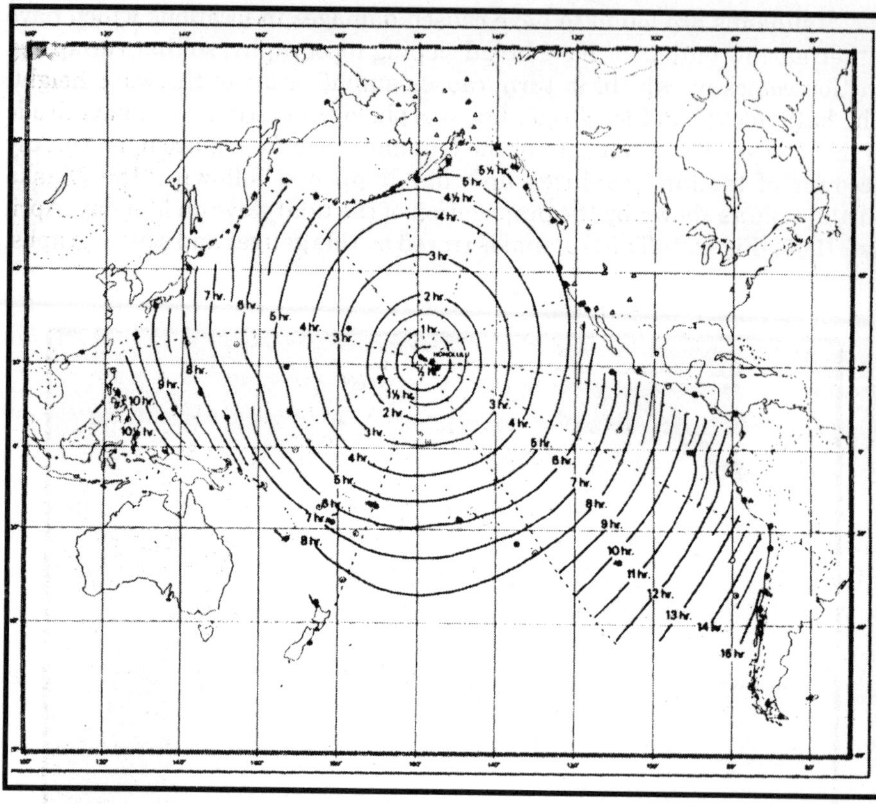

Fig. 2.3 : Tsunami travel time chart. Using instruments such as seismographs and tide gauges, scientists can forecast when a tsunami will arrive a given coast (Courtesy : UNESCO)

There may be a confusion as to whether this was a first tsunami in the Indian ocean. It is due to lack of proper documentation ; or do we have limited data in regard to tsunamis in the Indian ocean or is it that earlier occurrence(s) of a tsunami in the Indian ocean have gone unrecorded ? Specialists do not shy away from informing the people that the Indian ocean is vulnerable to tremors and so, why not tsunamis ? They talk about the sea-bed that runs along Australia, Indonesia and the Indian subcontinent and refer to this as Indo-Australian Plate. The early forties seem to have witnessed the so called tsunamis in Andaman-Nicobar islands, without any basis on records.

It is now a reality that the recent tsunami had its origin from the Sumatra earthquake in Indonesia. The sub-oceanic earthquake of Sumatra has reportedly brought about some extra-ordinary changes in the Indian lithospheric plate so that around 1200 kms of fault line slipped about 15 mts along the subduction zone where Indian plate drives under the Myanmar (Burma) plate. The Indira Point which forms India's southernmost edge has been obliterated from the Indian map due to the earthquake. According to experts abroad, "the earthquake occurred due to the massive amount of energy exerted and the sudden shift in mass" vide US Geophysical Survey, There are now many scientific findings on the Sumatra disaster.

Let us turn to Andaman-Nicobar islands, some of which are about 150 kms off North-western Sumatra. An aerial survey showed that one may have been cut in two parts and another divided into three parts. The island inhabited by Sentenelese is reported to have been restructured. As media say, a long stretch of coral reef, used to be under the sea, has, now, come above the sea and the islands seem to have tilted one side above the sea and the other below the water, but no one can say that whether it is due to the tsunamis or not and whether such changes are permanent or otherwise.

There were many tsunami scenarios which have been studied in depth, coupled with people's accounts and observations.

What we need now is to determine tsunami vulnerability in the Indian ocean. It is only when tsunami indices are identified, formulated, determined and prevailed upon, one can look forward to tackle the potential for producing the tsunamigenic seismicity in the Indian ocean.

Issues on Tsunami in the Indian Ocean : Looking Ahead

Having described the critical elements of conceptualising a tsunami, it would be, in the fitness of things, to have in view the issues that this tsunami has brought up. This is the other way of expressing what lessons ought to be learnt from the recent tsunamis. Let us proceed to do this, against the perspective that these problems and issues have many elements in common with those of other natural disasters. These are done in few sequences, as warranted by intrinsicalities and offshoots of issues concerned.

Lack of communication and information have come up to be a bane around reporting on tsunamis. Obviously, what is being prioritized among the desiderata is to go in for an investigating system directed towards reducing coastal vulnerability. Information Technology (IT), however powerful it may be, may not lead to us anywhere unless it is grounded on a solid foundation. Modelling seems to a panacea. This modelling ought to provide a design of a system which covers the wide spectrum beginning with problems of risks, rising from the oceanic seismicity followed by tsunamis, to the human system of community infrastructure and response. The three basic rudiments should be so structured that they facilitate in forging linkages between technology with absorption and adaptation geared to learning, problem solving and necessary implementation. In particular, IT should allow the decision-processes in inter-organisational networks by mapping the domain of choice opportunities on to the range of decision structure ; mapping problems onto respective choice of opportunities should also be an ally in this exercise. There is a knowledge base for the system in that it is cumulatively accessible to pro-active elements. A leading element should be to provide the critical measure of the outcomes of performances in response to the tsunami threat, regionally and in the global reduction of losses due to tsunamicity. Any exercise of modelling must involve possible players of the system which has to have alignments with community organisations, state organisations, national organisations and institutional organisations. The participation of these organisations should be transparent, at least in their respective functions, on disaster operations. One can delve deeper if we require the modelling to reflect a set of operating premises allowing interactions of the separate phases and levels, with the least of errors. Minimising errors can be sought in a variety of ways, for example, by standardized knowledge, a holistic perspective and the degree of complexity in operations. There is always a set of expected outcomes from such operations and these include inter–and intra-bases ; wider awareness of seismic and tsunamic risks along with alternatives for actions ; one expects most likely the possibility of timely and organised action for reducing as well, tsunamic risks. One cannot rule out flexibility on a wider scale in responding to such disastrous events. Efficiency in disaster operations is always sought after and so a reduction of losses in lives and property in tsunami-prone areas needs to be identified.

This model can, in no way, be full-proof unless assessed in terms of field observations. Such an exercise does exist for various case studies on disasters but treading the same path for tsunami-disaster operation is yet to be on the anvil. We need to take up some of the leading six to eight case studies on tsunamis which have witnessed the impact of IT on the organisational activities ; of course, occurrences of associated earthquakes, if any, automatically will come up. Hopefully, this venture will be revealing in many ways.

Thus, given the basic requirements regarding IT (Information Technology) for its linkages with organisational problems, one can develop a prototype model for the reduction of tsunamic risks. Expectedly, such a prototype might build a set of parallel technical and organisational systems geared to augment the capacity of communities for solving problems relating to tsunami risks and vulnerabilities. The timely feedback processes from leading cases of tsunamis, is a model that can be unidimensional, to begin with, which will impinge on other dimensions dependent on responses to assimilation and absorption and so, adaptation.

A contemporary trend, in keeping with the emergence of domineering market forces, is to be innovative about financing risks, involved disasters. Even the tsunamis in history haven't gone unheedeed while considering seriously the role of risk control and insurance in the earliest phase of disaster reduction. The concept of 'megaperils' has emerged because of risk vis a vis insurance concerns, for the future, although insurance has been, historically, one of the oldest strategies for coping with risks. Hence, the consideration of re-insurance. One has appropriately a poser in this context : can the risk control, an insurance industry, prepare for the impacts of megaperils (*megaperils* being defined as natural hazards with impacts producing damage of billions of dollars and going well beyond the boundaries of a single continent)? Obviously, a tsunami, particularly the recent one, has to be seen as an interesting exemplar of a megaperil. Tsunami is yet to be an insured risk ; therefore, there is a reasonable likelihood of a tendency to identify a disaster with the earthquake that triggered the tsunamis and so, one can tend to a kind of doctrine of approximate cause. A tsunami in the Indian ocean, it is to be noted, is independent of a similar historic event in the Pacific and vice versa. Further, that probabilities in a global market are additive has to be reckoned. As the economic world, in the Indian context, seems to be wide awake, one can follow a precept that policy holders are so paid in full, whatever be the terms of the policies. Tsunamis being megaperils, we need to address to ourselves such questions, recognizing certainly the complexity of natural perils. Honestly speaking and seriously as well, in regard to risk control, we need to know what produces damage. Moreover, what is it that produces insured losses? Obviously, these will raise many dilemmas if we seek to decide about policies in many markets. The reality is that little effort has been made to understand the geological and meteorological use of the term. The duration of the disaster is, in no way, purely of ritualistic use. Some policies can cover droughts, but not tsunamis and vice versa. Don't these land us in an arena of vexed issues when both belong to a continuum of disasters ? Megaperils, like tsunamis, thus provide good opportunities for risk control through well structured policy statements. Some one may cynically look at the industry, roping in lawyers for deep market syndrome. Are megaperils or natural perils temporarily random ? What can be the magnitude and frequency of tsunamis influenced by other factors ? Can we think of earthquakes along the tectonic Burma plate boundaries random or should we believe in seismic gaps ? Don't we come back to the earlier poser : are insured events independent of one another ?

3

PROFESSIONALIZING DISASTER MANAGEMENT

Introduction

Management of disasters is a fairly recent coinage. Before we seek the reason for using the term "managing" in the context of disasters, let us agree on the use word "disaster" with its several connotations. A disaster implies or connotes something which is sudden, unscheduled or perhaps unprecedented. That the event requires immediate action is also implied by use of the term disaster. Whatever be the urgency for such actions, required by its enormity and/or severity, there exists the use of another closely allied word "emergency" which, by and large, expresses when a disaster strikes. Strictly speaking, there is a distinct shred of meaning between disaster and emergency. Emergency is really a pro-active measure after the happening. Therefore, the management of either disaster or emergency, has to have separate characteristics, of course, with few overlaps. In what follows, we take up features that characterise management of disasters or emergencies with appropriate subtleties, if so required.

Disaster Management: a Prelude

The definition of a disaster can be well taken for granted, in that it is an event that brings in its trail several impacts such as on lives, properties, lifelines etc. Its management can be very naive to the extent we think about the immediate ways and means to stand up to it. One may tend to graft into it techniques and methods that rudiments of managing enterprises may provide.

It has been acknowledged that the disaster management is nothing sort of being a dynamic process in which the community participates along with several organisations. The ambit of such an endeavour is geared to work primarily on the adverse effects of a disaster. The dynamics of the ambit of the disaster management has preparedness, mitigation, relief, recovery and rehabilitation as its principal elements. The contours of disaster management should necessarily have several sub-areas of management depending basically on the phases of a disaster.

While proceeding with such exercises on discernible scores, the abiding but allied aspects of risks and vulnerabilities can in no way be ignored. Risk management, by itself, is fast assuming evolutionary traits because of anticipatory, acceptance, estimation characteristics etc. of a risk per se. Vulnerability management cannot obviously lag behind on account of the fundamental equation : Risk=Vulnerability+Hazard. Such an approach has become all the more important because of emphasis being stressed by the World Conference on Disaster Reduction (WCDR) held from January 18-22, 2005, at Kobe, Japan (also called the Hyogo Declaration).

Whatever be the complexion, the management of hazards acquires a pre-dominant character. Let us first dwell on this in general terms. Obviously key areas within the fold of hazard management should consist of (a) principles (b) practices (c) counter-disaster measures (d) other aspects conducive to socio-economic and cultural facets of the community concerned. In any form of practices on hazard management, the main rationale should be to look for ways and means for disaster reduction or even elimination. Sensitising the community on the cost and benefit, dominantly, human, can emerge from the hazard management. Developing public awareness and influencing public perception hazard, should be a major objective of anything to do with management of hazards ; hazard-specific managment practices should be identified. A vital element of this kind of management should be how to go in for containing or in a way, altering hazards. Steps in this direction ought to aim at providing the takers' ability for acquisition of knowledge with wider horizons. Appropriately, strategies on any front of hazard management can hardly go by default, for familiarity with such strategies will be enabling mechanisms for the purpose.

Hazard analysis, therefore, comes up as an important pre-requisite for any person who would like to handle hazards. The key elements of hazard analysis may apparently sound to be repetitive but the analysis always provides essential information required to develop and to act on preparedness, prevention and recovery measures. Hazard analysis can then deal with characteristics of a hazard ; then, follows the role of the community leading to what may be called community analysis. Perception of hazards vis a vis risk is always implicit in characterising hazards and analysing, allied aspects, even the community. One has to have necessarily, as part of the exercise of analysis, fairly explicit evaluation of hazards for which techniques and procedure should be reliable and credible. The element(s) of communication should be well passed on to the threatened community ; there should also be efforts to develop concepts and parameters of risk and vulnerability for those who are entrusted with the task of managing hazards. One can go in for a bit indepth for hazard analysis by investigating the processes of a analysis ; obviously without this exercise, there is likelihood that strategies on preparedness, mitigation, response and recovery may not be worth the purpose. Hazard characterisation, as a second step, should be in greater depth. Improved technologies such as those of information technology, satellite-remote sensing technology etc. being available now, there should be endeavours for holding programmes for the disaster management to have well-structured exposures. Risks and vulnerabilities can only be there when appropriate quantitative techniques such as probability, numerical weather prediction etc. be made components of equipmemt and knowledge of disaster managers. Any disaster management cannot, as an immediate measure, lose sight of problems of evacuation. There are, of course, some policy decisions about it but any venture on evacuation should aim at considering fairly deep perspectives of the problems and techniques of evacuation for communities in disaster.

Evacuation by itself may often appear to be a ritual ; this may be of use in some occasions but may turn out to be less than what it should have been, had very important concommitant aspects been reckoned. If an exercise on sheltering is resorted to without considering the nature of the community to be evacuated, in the event of a disaster, then it might be somewhat futile. What one should be equipped with is a comprehensive view of the total range of problems and methods of evacuation. The real wherewithal lies somewhere in the whole gamut of warning, recovery, sheltering, withdrawing and ultimately, the return of evacuees. It enables the concerned leadership to identify the components of evacuation and to proceed with decision on making processes. However nominal they may appear to be, there exists value oriented steps which are to be undertaken with utmost sincerity, sensibility and care. They may appear, even though marginal, pitfalls in respect of managing transport used for the affected community but these need to be attended to on ethical and often, on legal considerations. The behavorial aspects behind such processes can hardly be overlooked.

In every sector of disaster management what is often missing, possibly because of worse surroundings, is the human face. The initial step should, doubtless, be response to possible impacts of a hazard ; any worthwhile response requires understanding of the likely effects of the event, the dire needs of the affected people and the ability to harness the relevant resources so that managing disaster responses may not be exercises in futility. A disaster manager has to acquire the ability for appreciating a need for sharing venture with a view to have well–co-ordinated, well-structured and sustainable endeavours. Prioritising operational tasks in regard to decision making processes is a dominant ingredient of the whole spectrum of disaster response management. What often goes by default and allows people to be talking glibly is the transmission of right information at the right time to the people at disaster. It would be fair in the light of such exercises to perform what are often called, in the parlance of media, briefings and debriefings. An effective response management should, able to identify the discernible patterns of human behaviour of not merely be the victims but also of the workers during the disaster so that the camaraderie, more than a liason, is established on a long term basis. All these exercises can be, when added up, must depend on planning not only well in advance but well linked to the response. While management takes care of information and resources, pertaining to response, the relationship between response and recovery should kept in view. No management of this kind can be fruitful if the manager is not concerned with assessing risks and vulnerabilities associated with disasters. In terms of the current trends of the day, disaster management has acquired a place in the annals of the mission on the "knowledge", with explicit intervention by response agencies.

As a part of planning on disaster management, strategies cannot go by default; however, there may be apparently non-prominent incidents. The main purpose of disaster planning is to understand key issues during the process. These issues should necessarily involve the community and develop counter disaster plans within their own.

If the planning is to go wholehog, then one has to develop counter disaster planning models which, in turn, should depend on counter disaster planning concepts and processes. Such exercises should in no way be esoteric in nature ; on the other hand, planning problem solutions should have practical implications.

In brief, all the phases of a disaster should be well reflected in the planning processes of the disasters ; this means that preparedness, relief, mitigation and rehabilitation including recovery should be the foci of an effective disaster management. Areas such as risk assessment in planning, strategies for public awareness and training in disaster planning should be well interwoven with the above foci. It is not that, participation of usual agencies and organisations will ensure response to contend with disasters, planning should see to it that capabilities of intervening organisations should be improved.

Strategising Disaster Management : The Indian Scenario

It hardly requires any repetition that disaster management should essentially be a three pronged exercise, the first one at the pre-disaster stage, the second, during the disaster and the third, after the disaster. The management in the first step, is really to strengthen the level of preparedness and so, it is an attempt to encounter the impact, so that mitigation begins ; the second step is to mobilise efforts for shelter and rescue, of course, with relief being available by now ; the final, although a long term, one aims at another variant of three r's, namely, recovery, rehabilitation and renovation.

The Indian setting, whatever might have been its disabilities earlier, has seen fairly improved enterprises on disaster management. The focus of management, until recent years, had been purely on relief and rescue operations during disaster phase ; there has been no doubt some exercises on post-disaster rehabilitation. One can say now and more so, after IDNDR, that there is a greater awareness now of the importance of pre-disaster preparedness. Of late, preparedness has been fairly well professionalised and conceptualised, too. Who could envisage earlier the preparedness stage consisting of constructing physical infrastructures, roads, disaster resistant houses, community shelters, forecasts and warnings in coastal arena, at least for cyclones, storms, floods etc. ? Relief agencies, government or non-government, country based or global, have made tremendous strides in this direction. As an example, one can recall the devastating cyclone in Andhra Pradesh in 1969, which prompted the government to set up a Cyclone Distress Mitigation Committee in 1971, followed by Cyclone Review Committee by the Department of Science and Technology, Govt. of India. There used to be programmes on this theme between countries on a bilateral basis, for example, Indo-US workshop on Wind Storms. As a spin off of the recommendations of such deliberations, a network of cyclone warning radars was set up all along the coast. As a part of the activity of the management in pre-disaster phase, a well laid out plan is being prepared for any eventuality of this kind, with the sole purpose for disseminating the "warning" to the community. Evacuation ensues closely on the heels of this so that the subsequent steps by the government and others are well facilitated. The management should see to it that the evacuation strategy is well organised in that, its inherent limitations do not come in the way of carrying the affected community with them. Another example, may be about floods, for which flood plain zoning is well planned and monitored as a vital element of exercises on preparedness management. Although people, in general, are fairly well sensitized about its importance, several pressures and lobbying, mostly on commercial grounds, implementation of plans for floodplain zoning goes by default.

Likewise, in the case of earthquakes, hazard zoning is to be toned up as the foremost requirement for mitigating impacts of earthquakes. One still remembers the ravages caused by earthquakes in Latur and Bhuj. A wiser step should be to go in for micro-level zoning rather than macro-level zoning. The recent tsunami points to the urgency of steps, in regard to mapping of seismicity, whether they occur below the earth's crust or the sea level. The Government of India has set up a network of stations for evaluating seismic risks and vulnerabilities.

All these years the Ministry of Agriculture, Government of India, has been looking after disasters, till the Bhuj earthquake in 2001, when the onus has shifted to Home Ministry except for droughts, still being looked after by the Ministry of Agriculture. Several Contingency Action Plans for earthquakes, cyclones, floods, droughts etc. setting up of Calamity Relief

Fund, used to be the functions of National Advisory Council in the context of IDNDR. What ought to be in the offing is how to launch programmes, so as to ameliorate the impact of natural disasters and to equip the community with traits of resilience, so that challenges of natural disasters can be withstood.

In short, the disaster management in India, dominantly under public systems, has primarily reflected four dimensions : (i) use of science and technology for forecasts and warnings (ii) timely and orderly response (iii) mobilising community for participation, affected and non-affected (iv) equipping people with qualities of resilience for restoration of life styles.

In a country like ours, the overriding need is to build up capacities of societies, so as to meet the challenges of disasters. It looks that the recognition of this need has been fairly articulated but only reflected through relief efforts, that seem to be characterised by aspects of alleviation and to certain extent, amelioration only.

Ours is a country that cannot go back on processes of planning and, therefore, what we are in urgent need is to have a perspective view and action too, so that, vulnerability of disaster-prone areas is continually assessed. This does not, negatively speaking, in any way run counter to global trends, as are well set forth by the Hyogo declaration, Kobe, 2005. This declaration has stressed on disaster reduction through developmental efforts. Such efforts require necessarily a multipronged approach. Mitigation of disasters can be coupled with processes of development. As a part of the developmental process, an emerging area of concern is to put an emphasis on generation of employment, particularly, in rural areas, where socio-economically disabled groups carry on somehow, their livelihoods, stricken with worst forms of poverty. There are, no doubt, that the several programmes have been or are being launched for enhancing income of the poor families. So here ia an arena for alleviation of proverty which needs to be coupled with alleviation of conditions because of threats from natural disasters. WCDR has identified the decade 2005-2015 which coincides with the decade for providing water to areas of its scarcity and so, policy maker's concern ought to be directed towards provision of safe water for drinking in rural areas.

It is true that all the five year plans have sought to extend the areas under irrigated agriculture. What do relief operations on drought bring out the realities about the employment generation ? What are relevant outlays on this score ? What are their outcomes ? Will there be a match between outlays and outcomes in the context of allocation in the plan on assurance of employment for each family ? These will assume the status of glaring posers in contemporary times. For this, we need to recast our policies on natural disasters. Coming back to drought-disasters, it may be held that we reconceptualise "drought proofing" under Drought Prone Area Programme (DPAP), Desert Development Programme (DDP) and Watershed

Development Programme (WDP). DPAP, DDP and WDP are often made to converge into an integrated development programme without looking appropriately into their potential areas of convergence or divergence and so arise departures in respect of trajectories of each of them leading us frequently nowhere that is without any shred of convergence. Developmental processes thus suffer of not being sustainable and more so, in the context of natural disasters, vigorously emphasise the Hyogo Declaration, Kobe, 2005. Human development finds a marginal context in such processes and so the community response to natural disasters should acquire a built-in core of strength so that economic development can be brought to bear on the extent of losses of lives and properties during disasters. Developmental objectives should reckon disaster preparedness and mitigation, thereby meeting what are being envisaged through WCDR. One can move on in this direction in detailed aspects about confronting impending disasters. Organisationally speaking, an inkling has been put forward about what can

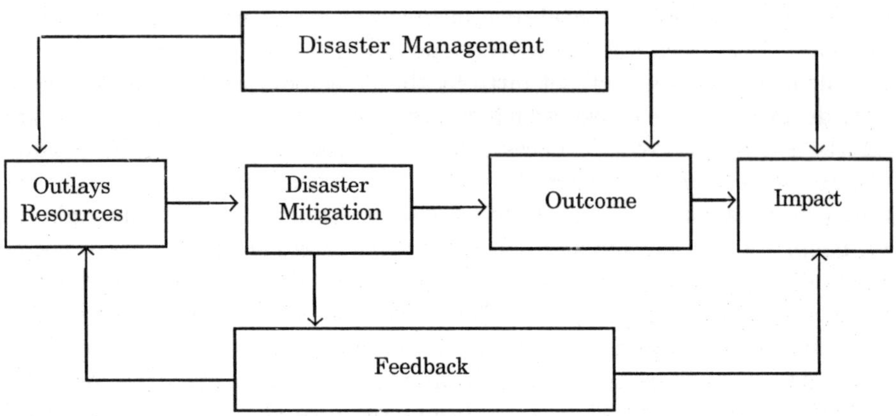

be done by a centralised agency, for example, the nodal Ministry of the Government of India. The State Governments are left to formulate contingency plans at the state and district levels, while there may be overarching frameworks for central and state levels, Such contingency plans can have a grassroot flair if Panchayet systems are well contained. Such exercises have to have programmes on awareness and training at all the discernibly district levels. All the steps for management plans can be well taken if there be a systemic view as shown by the above schematic diagram.

This view enables us to understand the planning and management process in its totality and empowers disaster management personnel to have the perspective view as well.

Remark

The foregoing lines, doubtless, indicate in fairly expressive terms that disaster management is a multidisciplinary and multiagency endeavour, aiming at managing the situation that would be necessary during emergency. Needless to add that emergencies have some distinguishably characteristic features, which need quick decisions under trying circumstances. This situation can hardly make strides if there be neither personnel nor materials. The chapter on emergency management is, therefore, intended as a sequel to disaster management.

4

ON THE RECENT TSUNAMI: AN ANALYTIC APPROACH

Introduction

The recent tsunami triggered by the Sumatra seismicity, has assumed an extra-ordinarily much -hyped complexion. It is not just simply the phenomenon on December 26, 2004 ; the more contemporary earthquake on March 28, 2005, has not only reinforced the concern about tsunamis, but also raked up a variety of issues, some of which, could not have been presumably attended to earlier, far from being conceptualised. It has so happened, since the first occurrence of a natural disaster, that myths or rumours spread around it and the same seems to be the case with the recent tsunami. Whatever might have been its genesis, and its aftermath, of which people, the media and policymakers often glibly speak about it, the ground reality is that this event, namely, the tsunami, has to be looked at, from a point of view that dosen't have any parallel in the history. The March 28th 2005, seismicity did legitimately raise alarms which came out to be somewhat false, as have occurred often with the warning systems at Honolulu (Hawaii) in the Pacific. One finds accusations of one against the other, on the basis of hearsays or untruths that go around in the wake of a disaster. Nevertheless, one could find some criticalities in respect of observations from various sectors of the society who rose to the occasion. Even the fishermen in the shores and indigeneous people in the islands could respond effectively on the basis of their time-honoured experiences through generations, whatever be the shortcomings for rescuing or providing shelters. There is, no doubt, an enormous spate of furore in a variety of ways from which, positively speaking, one can learn lessons from the recent tsunamis. Some leading posers in regard to the handling of this tsunami have been identified and their answers, provide, in some ways, sustainable indicators for the future. Only a critique of such developments and responses can bring out what should be in store in regard to managing future disasters of this kind. This is precisely the purpose of this chapter. That an analytic approach, without being bereft of scientific temper and tenor on such occasions is called for, has been sought in this chapter. Further, the human element, which plays a vital role for future strategies, figures prominently in this presentation. That there exist socio-economic realities and futurities as well, are also considered here and more so, in the context of overarching globality of issues and concerns and underlying vulnerability of the arena concerned.

Tsunami–Some Preliminary Remarks

There has hardly been, by now, any preliminary introduction about tsunamis without its etymological connotations. The terms 'tidal wave' are in vogue, in popular parlance so as to speak of happenings on level changes in oceans. It is given a technical designation, borrowed from the Japanese word tsunami (*tsu-nah-mi*, equal accent on all syllables, sneeze the first one) ; *tsu* means **harbour** and *nami*, **wave**. So tsunami represents essentially a series of waves in the ocean generated thereby a massive earth movement process on seabed, such as an earthquake or a volcanic eruption or extensive landslides. It's fairly well known that all land earthquakes are followed by surface fault movements and so, likewise, not all sub-oceanic shocks are accompanied by changes in the sea-floor resulting in tsunamis. The recent two events on December 26, 2004 and March 28, 2005 around Sumatra bear well testimony to this. One can cite copious statistical figures to indicate that only a very small percentage of oceanic earthquakes can bring about destructive tsunami (s). The most notable feature of a tsunami is that its first motion along a shoreline is a recession of the water as if by an abnormally low tide. But this is really the forerunner of large waves that the inhabitants living in places prone to these phenomena, have learnt to have the warning about the tsunami and so, the people get time to flee to higher grounds. Haven't we heard about it from what Nícobaris, Shompeneses, Jaroas and others in Andaman and Nicobar islands did on their own? Out in the ocean, the wave is deceptively not more than a metre high and so is rarely visible. The are several instances where the sea recorded several kilometers, although the inrush, as a reaction, sometime follows quickly the withdrawal. The time for the water' s return is somewhat variable. The height of the returning wave depends on the nature and the depth of the bottom at any point. It is seen that over a shallow-bottom or in a near bay (if, in particular, it points to the wave source) waves keep on piling up to considerable heights. Indeed, by the time it approaches the coast, it has grown enormously, may be 25 metres high. Tsunamis occur frequently around the Pacific, travel very rapidly and bring about alarmingly terrific damages. For example in May 1960, the very violent earthquake in Chile created a tsunami which caused hundreds of deaths in Japan, few in the Hawaiian islands and subsequently dazed the largest statue site on Easter Islands. In fact, the Chilean tsunami originating from of Southern Chile struck the coast of Japan from Hokaido to the Ryukuu islands. Although it reached the Japanese coast 22 to 24 hours after the earthquake, no warning could be given before the first arrival. The maximum wave height exceeded six metres in some bays of Sanri coast. As figures go, 139 people died, 872 injured, 2830 houses and 1500 ships destroyed. The violent currents in Kiipeninsula very close to Shima, fish farming also suffered a lot. In the case of tsunamis generated in Japan, the first arrival of the sea waves being 10 to 30 minutes after an earthquake, it was possible to predict its occurrence. Indeed, in the coastal region, the tsunami disasters occur repeatedly, people become experienced how to reduce the risk in tsunami disasters. The emergency-measure requires that people should immediately move to high places, although relocation of houses to higher places, building up of seawalls, tidewater control, afforestation and shelter harbours are essential in tsunami-striken areas.

Distribution of wave heights on the coastal areas differ from one tsunami to the other. It is believed that V-shaped narrow-based provide conditions favouring the increase of wave height towards the bay bottom. The increase in wave height occurs when the energy of the tsunamis converges to the sciche of the bay. There are usually low land areas at the head of embayments where one finds a motley of settlements, mostly the fishermen.

There are two Regional Tsunami Warning Centres established in the Pacific–one in Alaska and the other, in Hawaii. These Centres deal with tsunami warnings in northern part of the Pacific. The mid-sixties UNESCO Inter-governmental Oceanographic Commission (IOC) set up an International Tsunami Information Centre (ITIC) in Honolulu which works closely with Tsunami Warning System in the Pacific operated by USA. ITIC provides support for establishing national tsunami warning systems and extends technical knowledge on the latest warning system equipment and survey procedures. It has obviously a computerised storage and retrieval system on tsunamic data. Tsunami warning covers all coastal areas of Japan and depends on ocean wave advisories and warnings, and also marine meteorological forecasts and warnings on Earthquake Phenomena Observation System. The equipment consists of strain meters, tilt-meters, extensometers, ground water level meters, tidal guage etc. Warnings are routed through Information Transmission Centres, local meteorological observatories and other mass media; local establishments are obviously sent through TV and radio stations.

Tsunami Studies : A Critique

In Japan, there are several studies on tsunamis being generated by submarine landslides without being accompanied by so called foreshocks, as in the case of earthquakes and volcanic eruptions, obviously predicting wave heights of various dimensions. Japan, needless to add, has witnessed many tsunamis and, in fact, has an edge over others. Some of these tsunamis, in a big way, provide the basis for the development of modern tsunami science and technology. For example, the Nihonki-Chubu earthquake tsunami of 1983 has given rise to a variety of tsunami data, while seismic data are available from land-based seismic observation stations and a temporary Ocean Bottom Seismogaph (OBS) array. For quantitative analysis, tsunami data consist of tide records and runup heights measured in extensive surveys; for purposes of qualitative analysis, numerous photos and videos are used. There are several model studies in this direction. One often associates the total tsunami energy with the fault parmeters ; the same method applies with tsunami runup heights, but the vertical sea bottom displacements are often left unexplained. The relevant model study, after modifying the sea-bottom displacement satisfying the total tsunami energy, could compute the maximum height. There are studies through simulation where the height of the first wave could be obtained. The arrival time could also be obtained along with tsunami propagation profiles. The 1964 Great Alaska earthquake, when analysed, did imply a lot about the nature of shocks. One finds often contradictions between seismic manifestations and tsunami phenomena. In fact, tsunami simulation continues to be a challenging problem as to whether or not the small first wave yields the higher runup. One might tend to think in this direction in respect of what we find in Andaman & Nicobars wreaked by recent tsunamis. There ought to be a modern tchnology for hindcasting of the tsunamis. In sum, numerical models should be capable of reproducing the basic features of tsunami phenomena, as against by computer simulation of many historical tsunamis. There are pursuits in this direction. The study of 'design-tsunamis' has already come up, specially in Japan and Russian Federation. In the erstwhile Soviet Union, there were studies on the abnormally high tsunami waves caused by relatively small magnitudes of earthquakes. There are evidences that the 'runup height' depends largely on the shape of the coming wave. In the Institute of Oceanology, Moscow, there are many model studies on the generation of tsunamis. There are endeavours on tsunami generation in the same Institute comparing energies of earthquakes and tsunamis. In the Russian Federation, they were looking for information on the initial tsunami source because efforts, for operative tsunami prognosis, become somewhat inhibited on account of the lack of time for issuing of tsunami warnings and the lack of computational warnings in the existing warning centres. Indeed, the Interactive Tsunami Modelling System (ITMS) for numerical simulation of regional tsunamis has been developed at Novosibirsk Computing Center. There are also simulation studies on generation and propagation in the ocean with real bathymetry within the nonlinear shallow water model of hydrodynamics. Tsunamis occur in the Mediterranean area as shown by scientists belonging to Institute

Nationale de Geofisica, Rome, Italy. Sicily, replete with volcanic eruptions, has witnessed tsunamis of great diversities and adversities. The tsunami runup, being the most important factor for warning and designing, any protection work is being studied in depth by the Japanese scientists at Kyoto University.

Some Aspects of Tsunami Issues

Tsunamis are relatively rare hazards, but as recent events show, tsunamis are capable of causing hundreds of thousands of fatalities and bringing about disastrous damages both near the source and at tele-tsunami distances. Obviously, one has to works strategies to deal with problem for at the local level, whether tsunami source is local or remote. One has to have a knowledge of the local tsunami history, nature of inundation and currents, geophysical results of strain accumulations, effects of the guage ranges etc. What has happened in recent tsunamis, poses a variety of questions : Where was the source ? What was the cause of the tsunami ? If local, as in Andaman & Nicobars, how much time was there between the generation and arrival of the first wave ? What was the time of day as well as the state of tide ? Did the first wave show a rise or a fall (recession) ? Which wave was the largest ? What can be said about the nature of the wave ? Was it a bore, breaking wave or a flood ? What were the effects of water currents? What was the extent of damage ? What were the circumstances under which news on casualties keep on emanating ? What was the maximum wave or runup height ? One might go in for several case studies. What would have been the outcome if the policy makers had known the lessons of tsunamis beforehand ? Do tsunami lessons enable policy makers to develop strategies for preparedness and mitigation tsunamis ? How can we situate 'tsunamis' into the broader international context, by concentrating the Andaman-Nicobar case on the basis of approaches being taken elsewhere ? Shouldn't now there be pursuits to develop baseline information on social, economic, political and demographic characteristics that would be useful in assessing the impacts of tsunamis in the southern part of India and Andaman & Nicobar islands? Isn't it that the information has not been obtained so as to include risk perceptions or organisational response to emerging situations ? It looks that information base should include much more data on institutional and agricultural phenomena, rather than simply inventories on equipment, emergency facilities or the like.

The sequence of events in regard to the tsunami phenomenon, being the first if its kind , the recovery process has to be a long-term affair. Few fundamental questions are still raised. What is recovery and how does the recovery occur ? Both the definition and the outcome of the recovery are largely determined by linking, in particular, social, economic, cultural and political contexts ; of course, the human context is the overriding one, shaping, how we are used to look at the recovery and also the people who are going to recover.

Lifelines are so complex and diverse, that one ought to be sure, about what are precisely meant by 'lifelines' and in particular, lifeline hardware – their functions and conditions in the context of damages. These are well needed with distinct uniform database. Damage aspects ought to be defined in that they relate to various lifelines which are, of course, true whatever be the type of the disaster. Utilities and lifelines should go together, specially in respect of the impact of the vulnerability and impact studies. As is often posed, do lifelines put up a threat on lives if they fail ? We did see this in recent tsunamis.

Some Lessons from the Recent Tsunami : A Prelude to Future Strategies

There is a lot of contrasts in the general public's view of the phenomenon with that of the technical community. Strategies should obviously seek for resolution of the technical, management and policy issues.

The impact of life-line failures in secondary and tertiary losses comes up as a major, if not common, concern among while investigating the effects of natural disasters on regional and national economics. Losses due to tsunamis waves, coming one after another through intervals, ought to be investigated, because of possible contrasting in losses, associated with an interruption of life-line services. There is decisively an indirect loss assessment which requires an interdisciplinary endeavour, particularly by engineers and social scientists for the simple reason that technologists do not appreciate the nuances or dimensions used by social scientists for modelling socio-economic impacts ; the converse is true in that social scientists may not be able to use the language of engineers in respect of life-lines. The recent tsunamis should provide us with data that might affect the future of lifelines.

This assumes a distinctive character when we think about of tsunami mitigation. A "soft" mitigation is one of educational programmes rather informative dissemination programmes about it. But the information seeks basically, in many ways, reduction of tsunamis. Yet, a harder phase of mitigation measure ensures which one goes in for estimating the cost of mitigation measures(s). If such measures are to balance against payoff, we should go in for analysis based on more of the other aspects of costs of the efforts, for purposes of analysing cost effectiveness. A coastal process-perspective and fragility, in particular, should be to devise a new strategy to reduce damages for tsunamis. One should envisage, besides a conventional mitigation, options, new mitigation concepts because of arrival of tsunamis at the shore in intervals.

Various dimensions of recovery obviously warrant further consideration because of the character of the natural disaster(s) which we haven't seen earlier. These give rise to issues such as (a) how to find long-term recovery (b) mechanics for sharing lessons learnt from disasters. We cannot shy away from questions, relating to populations of Andaman-Nicobar islands, the special character of needs ; how to empower these special populations with capabilities and the backlash, as well. If we reckon futurity, we must see to it that there is effective pre-disaster planning for coping with tsunamis and certainly, of course, cost-effective mitigation measures. We talk about recovery and response to this slow on-going phenomena of tsunamis and also that, we should be having financial support of one kind or the other, at the global level. The planning on logistics becomes then an integral part of the initial planning process for any operation. Indeed, the flow of logistics has to be carefully structured so that accountability of this system is well taken care of. One can hardly dispense with competitions, now-a-days, between private and business in increased dimensions. One should look at afresh about the role of inter-organisational net-working disaster preparedness, and may be, from now, one may have to define what are meant by disaster-preparedness responsibilities.

One of the tragic parts of the tsunamis in the southern part of India is that so many children have become orphans. The realization is that there should be a full-time clinical consultant and trainer in disaster mental health. There should be exposures for children, school teachers and other personnel such as mental health professionals and others concerned with assisting children in their emotional recovery from disasters ; also, how to make children learn techniques and tools in the classroom scenarios and having known these, equipping them with the ability to refer a chiild for additional assistance or counseling, videotape may be envisaged for the purpose of empowering children and the school personnel. In brief, steps, on a national level, are called for educating children about disasters and preparation of age-appropriate materials and activities for the purpose. The details of curriculum development may be proposed for educating children and the family, as well, particularly for areas fraught with coastal vulnerability, pertaining to Indian ocean.

Tsunamis have brought to the fore the accounts of victims in low income groups of families. Despite all these disabilities, there are sparkling pockets of ethnicity whose possessors are remarkably well equipped with the wisdom on how to grapple with natural disasters like tsunamis. There exists, remarkably, a plethora of ethnic practices, customs and styles of living which ought to provide a repertoire of what may be called ethno-disasterology strategies. Further, despite fragilities of all kinds, resilience ought to assert itself organically, over and above, available relief, recovery and rehabilitation.

Tsunamis have given rise to not merely problems of great complexity but also surprisingly those that have implications with far-ranging economic and social connotations. The global view continues to be relentless. Recently in the Indian ocean, the coastal vulnerability stands terribly exposed, and worse than what happened to be when tsunami hit the Pacific, wreaking upon Chile, Japan etc. There is always a propensity in case of a disaster like this, to draw several analogies, even though commonalities can hardly be precluded. The event in Pacific didn't have, historically speaking, any precursor in the form of globally sponsored convention. The nineties, just gone by, did have several of them and served as for Preparedness, Mitigation and Prevention (PMP) strategies for several disasters, feeble or strong. The World Conference on Disaster Reduction (WCDR) held at Kobe in January 2005 is not purely a rehash of what IDNDR stressed earlier. The tsunami in December 2004 obviously prevailed upon the deliberations of WCDR in that the vulnerability *per se* could come up as an area of vital concern, particularly in regard to coastal regions. The absence of forecasts and warnings, because of the lack of an approximate warning system, was keenly felt. The poignancy of the tsunami devastation was not simply confined to limits of emotions and feelings. WCDR really plunged itself into deliberations that covered vulnerability in the widest possible arena. Spatial and temporal vulnerabilities were looked at anew because of what occurred distressingly in Sumatra, Andaman & Nicobars, Sri Lanka. Here was a tsunami which didn't limit its destructive characteristics to preferred communities; indeed, it was totally, unlike, again going back to the pages of history, what took place in Lisbon tsunami that led 'several people' to have a vivid delineation of vulnerable communities. Voltaire's Candide portrays the bathos of sufferings and angst stemming from what used to be labeled 'classquakes'. Poverty reduction finds a place in WCDR not as a purely leap service exercise, but as something adjunct to vulnerability. The development, or truly speaking, sustainable development, is often cited as a cliche in global summits, but sustainablity connotes something really significant, again within the purview of vulnerability. The potential hazards inevitably keep on lurking back in the minds of not only the sufferers but also the policy makers and custodians of PMP strategies. Doesn't it get us back to the abiding equation : risk=hazard+vulnerability ?

What matter most are areas where hazards of various kinds are known to occur and people chose to live there. Such areas, prone to hazards, are often areas of economic benefits. The people do perceive the risks and still, they decide to live in such areas so as to draw upon the economic advantages that the areas may have. The risk from the hazards is not often an overnight phenomenon ; disasters might have struck the area in the past and may well do so in the future. Hence, people move from the phase of perception to one of acceptable risks, after weighing the possible economic loss or benefit. Such decisions by the people do not easily emerge for communication to those in other parts of the world, as yet immune from disasters. People are now encountering a dilemma where the positive, economic effects and negative, disastrous effects coexist. Even though, we go by the adage "to err is human", here, people seem to be confronting with a kind of vulnerability, reflecting the degree to which communities, families, and populations are capable of adjusting to or absorbing relatively, the latter effects. In a way the more the vulnerability, the less is the capacity for absorption and adaptation. In fact, one can identify the core of vulnerability to be the cumulative effect of a variety of factors from which people can hardly make right choices. As already mentioned, poverty reduction has come up as an important area of concern in WCDR ; more so, in its trail, the poverty of education, information and access to the whole range of resources and support. Isn't it then, three A's – Affordability, Accessibility and Adaptability – should be a cardinal trio that has a domineering impact on the prevailing system and a wide range of inhibiting factors? Local resources such as education, sensitization, investment, access to markets and wide range of freedom do provide choice of the areas by the people. One shouldn't rule out some macro-factors resulting from proverty – population growth, urbanization and environmental degradation. Thus people keep on surviving somehow while struggling within a fragile economic framework dominated by low economic levels and livelihoods, irretrievably at risk. This may apparently sound to be the most seamy scenario at its worst form, but what else the impact of a disaster can be without being catalysmic, if the livelihood of people is conditioned by such negativities ? The tsunamis in the islands or coastal zone where people lived under economically throttling conditions, bear well testimony to the scenario described just. We seem to have reconciled to a reality that the recovery from the tsunami impacts will take a long time to come about. Indeed, recovery or rehabilitation from such an impact to the position of acceptable living standards, may turn out to be impossible so that the affected communities or segments of them may be left in the lurch and so, become more vulnerable to future disasters. This warrants a contextual model where we have to have basic presumption on the overall objective of disaster management without responding ritualistically to the kind of vulnerability, as cited above. In simple terms, the objective should be to hasten the process of minimization gradually, of the impact of disasters on population unless the factors which contribute to vulnerability to disasters are adroitly tackled, there is every likelihood that such factors will surreptiously creep into any form of emergency management aiming at something arbitrary and with adhoc-ism.

Model Approach : Its Raison d'etre and Salients

If we seek to conceptualise vulnerability pertaining to disasters, it will, as a consequence, lead us to a better understanding of the relationship between disasters and development. WCDR has stressed rightly on poverty reduction *vis a vis* development. It is undeniably true that there can be endeavours to kickstart development in the post-disaster phase. These development programmes are often formulated without undertaking appropriate cost-benefit analysis in terms of risks and vulnerabilities. It might so happen, if arbitrariness is allowed to hold sway that for the benefit of "evacuees", acquisition of land may result in "oustees" so that one sort of resettlement might give rise to a new phenomenon of resettlement. We may be drawn unknowingly into a complex web of problems of resettlement and rehabilitation. We have to have, necessarily, then, a model of risk analysis focusing on mitigating risks coupled with alternative livelihood strategies.

One does not know appropriately what ought to be the complexion of action plan on income generation strategies. Sustainable development can hardly be a reality because of such an improvement of risks involved. Truly speaking, one should go in for a theoretical model that conceptualizes sustainable development as the process consisting precisely of, in the context of disasters: (a) areas of shelter and rescue (b) the natural ambience (c) establishment (d) improved versions of technology (e) perception and acceptance of risks and vulnerabilities, and (f) three A's, as mentioned above. Here is a model that can be unidimensional, to begin with, which will impinge on other dimensions dependent on responses to assimilation, absorption and so, adaptation. The indigenous wisdom is often ignored in disaster management. How can one have a behavioural model so as to be of use in creating effective community response and also, sustainable ventures ? Can this model ill afford to ignore the complexity and interconnectivity in regard to a model on organisational development ? The recent tsunami calls for a fresh look at coastal vulnerability in terms of futurity, construction of mangroves and sea-barriers. One has, again, to turn sustainable forestry which is ecologically viable, economically feasible and socially desirable. Indigenous ethos of use to forest resources provide facilitators in this direction. One can cite a variety of examples from African subcontinent to Latin American countries to draw upon indigenous practices so as to reduce the risk that may be impoverished or terribly reinforced because of the intrusion of market liberalization, over and above, risks and vulnerabilities due to disasters. Indigenous cultures can always be rejuvenated for a sustainable feature in the context of global homogenization. A seriously missing element in the context of recent tsunamis in the Indian ocean, is that it has been in no way linked with climatic change and its possible impacts on some of the poorest and most vulnerable communities along the coasts, in future. Let us reconceptualise vulnerability in the context of a community. This is undoubtedly a function of climatic conditions and its adaptability to climatic hazards. A community's exposure to a hazardous environment mirrors not merely the occurrence of the disaster but also the location and structure of the community. The ability of a community to prepare itself for coping with or recovery from, exposure to disaster related risks is determined by characteristics of community and its vulnerabilities. A schematic diagram of the adaptive capacity to natural disasters is presented later in the next page.

Societal foci : Mesoscale standpoint

Let us dwell on some consequential societal implications. We begin with the notion of a mesoscale weather-related event pertaining to natural disasters. The societal considerations are warranted because of (i) people are ordinarily aware of weather events in their locales, (ii) it is the scale at which humans directly interact with atmospheric processes, and finally, apparently an esoteric one, (iii) mesoscale pursuits are more tractable than their counterparts in the temporal and spatial scales.

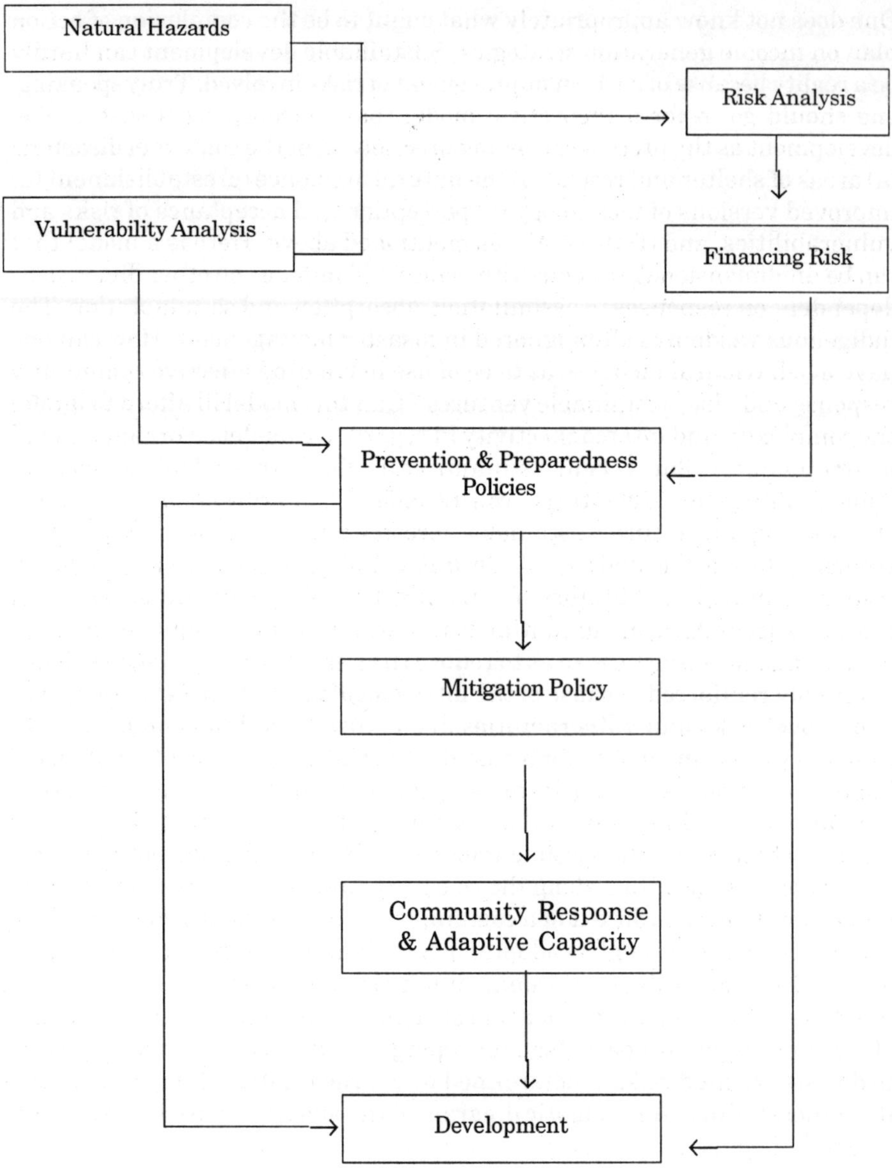

Having reckoned the rationale, we can proceed, as we do in the case of a disaster, with three phases, namely, (a) forecasts (b) impacts, and (c) responses. The first one is dependent on dissemination and the second one is an interactive phase relating to society and mesoscale weather events and finally, the third one, is nothing short of being a community response. The tsuanmi events, in recent times, call for endeavours well beyond what have happened namely assessment, coastal vulnerability and climatic change. Problems, questions and issues come up in the following simple forms : (a) what has been done in the area of tsunamis ? (b) what is not being done ? (c) what are the gaps in our knowledge ? (d) what are the issues of desiderata to be addressed in this area ? (e) what science and technology need to be generated to fill in the gaps ? Tsunamis, thus, offer in-depth deliberations and research, as well, particularly with a processual complexion : forecasting process, communication process, users' needs and decision processess. If we are effectively through such processess, we will be arriving at a situation that can be well integrated with global economy; rapid responses will be domineering features and mesoscale forecast will make efficient use of resources. We cannot have, without mesoscale impact, databases for the dire need for centralization and networking in the development phase. Tsunami requires invariably development, establishment and maintenance of communication network, so sorely missing in contemporary times and leaving so many lives and properties heavily devastated.

Winding Up

We see around us reports in the media, by the World Bank, Asian Development Bank and few allied bodies on the estimation of losses due to tsunamis in terms of billions of dollars. But where do the specific estimates stand ? If there apparently be no concerns of those who are commissioned to do so, how would public and private sector organizations would work for life safety? If some sort of prediction is obtainable from such exercises, it could be of use to improve natural hazard mapping and strategize risks and other enterprises on management. From the economic standpoint, there could thus be a spinoff for development of realistic scenarios for simulations; for planners to predict the impacts of hazards on vulnerable subpopulations, like the indigenous populations in Andaman & Nicobar islands and to allocate adequate resources for a variety of genres of natural calamities without being ritualistic about them. Economic consequences of natural disasters have, doubtless, variables depending on the nature of hazards, particularly their discernible characteristics and so, is warranted the economic analysis of disasters from a holistic standpoint. Neither we are fully aware about the vulnerability of key economic institutions. nor we have ever ventured to consider how they could be tailored to limit post-disaster repercussions. The new Agency, being envisaged at the central level will, hopefully, look into this. A report on the findings and conclusions from policy-relevant materials about tsunamis ought to be brought out. The materials collected for the report should include the information obtainable from idividuals or organizations with natural disaster expertise or some sort of official standing, provided the information has policy implications. These should necessarily focus on problems amd opportunities for purpose of dissemination, which may be done through NGOs and in-and non-formal outreaches. There ought to be, in the report, stuff for researchers and should provide enabling means to make progress being made with respect to policy changes.

One should wind up by stating that throughout the world, many issues often dominate understanding of "disaster management" policy-making at the national level. Broadly speaking, the inland issues are, by and large, fairly well contained but the recent tsunami has brought up coastal ones afresh. The climatic change will have, no doubt, implications. Given an effective warning system, questions arise : how would a manifest disaster worsen further ? Also, given the explicit expectation of a natural disaster, how can the policy-makers play their vital roles in not only understanding a creeping disaster but also move over from a static contingency framework to the required dynamics of the decision path ? This requires, irresistibly, a decision path model. In response to such posers, we may have often to fall back upon, hopelessly, on the conventionally outmoded ways of tackling the disasters. A child's wisdom (!) on informing his British elders that the recent waves on the Indian shores started receding continues to be an exemplar of disseminating warning, which may not even work, if there be a sudden (or persisting) disappearance of power and communication, undoing what the available technology could have somehow resolved.

5

MANAGING CRISIS

Introduction

In a previous chapter, we have mentioned about managing disasters and emergencies and also, touched upon cursorily, managing contingencies. These words, as mentioned earlier, are used because of their nuances in the context of specific scenarios. Let us now consider the term "crisis", without being a synonym of the others. The usage of the word crisis is mostly in government circles, although the word has several connotations well beyond its use by establishments. Without setting forth its formal definition, let us have an overview of its attributes, usages and rationale, too. Etymologically speaking, the word 'crisis' means to decide upon. But when ? The answer, even though implicit, is provided again by the contexts of its use. That, perhaps, is an enabler for a fairly straight-forward definition. Can we define crisis as a disruption that not merely impinges on the system holistically but puts its constituents in jeopardy ? Obviously, there is an element of threat to the system. As is obvious by now, there are interfaces with ingredients of a disaster.

Crisis Preliminaries

Like a disaster, there exists around a crisis, a variety of characteristics such as the "disruption" being un-scheduled, un-expected, un-planned, unpleasant, sometimes un-imaginable and often un-manageable and so there is, in a way, an halo of "un-ness" as, remarked by Hewitt. [11] We often associate some forms of critical episodes with crises.Crises are concrete events with a multiplicity of realities. Indeed, crisis reality is far from being homogeneous and therefore, we need to explore it in terms of a variety of realities. Isn't it because of multiplicity of crises, that there can hardly be convergences of perceptions and acceptance of crises ? Crises may be rapid onset, self-evident and acute but they may also be slow onset–creeping, emergent and ambivalent. We know of countries and regions that frequently undergo cyclones and earthquakes and so develop a kind of routinization ; presumably, we can call it a crisis (or disaster) sub-culture.

We have referred to threat as a significant element of crises with concomitant horrendous possibilities and experiences. Likewise, in disaster prone-areas a threat will be in no way less severe in disaster scenarios, bereft of massive destruction but which, instead, figure observable dangers. One cannot rule out disasters having no aftermath physically but leaving a devastation on the lifestyles of the community. Conceptualising a threat becomes an onerous task because of so many inherent and contentious conflicts and contradictions.

We know, in terms of conflict – resolution, that what apparently seems to be a threat, averting solution to a conflict problem, may turn out as a crisis-provoking step. It is there that the right crisis management fits in. Thus, the gamut streching from a severe threat to an affordable opportunity provides a continuum of endeavours by a crisis manager.

One of the rudiments in the "un-ness" is inevitably "un-certainty". Further, this is typically an "un-ambiguous" term. In disaster management, we take recourse to, in its very initial phase, exercises on shelter and rescue away from the place of occurrence of the event. Isn't it an effort to isolate the crisis or to cordon off as a no-go area ? Once this isolation of crisis is realised, one does necessarily reduce the level of misinformation which may increase the degree of uncertainty. The situation may worsen if there be a lack of appropriate media coverage. There may often be a lurking cynicism about prediction of crisis (or disaster). Any kind of vacillation or predictablity, from scientific standpoint, to an acceptable extent, may augment the level of uncertainty of the crisis. Since the community is involved, uncertainty often acquires a collective dimension which, in turn, may give rise to a stressful context.

That crises induce urgency goes without saying. It, in a way, being largely synonymous with emergency, assumes a compelling characteristic of crises. In any contextual scenario of a disaster, a crisis manager is required to intervene within a few hours, minutes or even a split of a second, so that the sheer existence of people may not be in danger. It is a pity that the sense of urgency is often being felt somewhat subjectively so that disasters can hardly be rolled back in regard to the impact on the people facing criticalities of helpless circumstances. That lack of action on urgencies may make communities fall prey to a fatal syndrome is well borne out by the recent tsunami in the Indian ocean. The lack on deciding upon emergency measures may explore itself through abilities of not posessing or even committing to costly ressources for an investment which the horrible in-situ predicament demands. Of course, such a situation should in no way foreclose the possibilities for future activities and programmes on this score.

In sum, to grapple with a crisis, we will have to look upon it as a dominant part of an continual process rather than an isolated event. If the processual nature of a crisis is accepted, its dynamics, both in retrospect and prospect, has to be well structured and monitored. Going back to the root-meaning of the word "crisis", we can ill afford to decide on fiascos that may ensue for lack of long term perspectives. Crises, whatever be their nature, rapid or slow, have trajectories with non-linearities and hence, phases of evolution.

Remark

With the foregoing introductory notions on crises, we proceed now with intricacies of crisis management involving some of its possible phases similar to those of natural disasters.

Rudiments of Crisis Management

As the words imply, any enterprise on crisis management should not conventionally leave an impression that given a problem, it would be possible to arrive always at its solution. For, problematic situations may not be amenable to easy or shortcut resolutions ; indeed, there are neither soft problems nor easy workable solutions.

Let us now distinguish what can form the contours of a possible crisis management system. It may be recalled that we have defined it on the basis of some processual elements. How can we explore them pertaining to disasters ? A simple approach should be to go ahead with a phase wise structure consisting of three 'p's viz. prevention, preparation, planning ; four 'r's viz. response, recovery, relief and rehabilitation. The sequence of these phases remains incomplete if mitigation as a distinct phase is not included. In order that the crisis management reflects broadly the bulk of reality, we have to have some other sub-phases that should indicate three c's ; consolidation, continuity and change ; for example, recovery does relate to all these three c's. Let us now identify the leading heights of the sequence of phases of crisis management:

1. Prevention : Clearly the focus should be on efforts which prevent crises from occurring : it assumes exposure to crisis – proof considerations, enterprises so that expenses on planning and preparatory programmes are not heavily incurred.

2. Preparation : The specifications of preparatory programmes depend largely on pre-crisis condition and its domain or activity should be undoubtedly resolute, which, in turn, is well reflected through programmes closely on the heels of this, namely, preventive efforts ; the other main focus of crisis management should be on planning.

3. Planning **: In planning, one has to keep in view scenarios covering the entire domain of risks, vulnerabilities, threats and devastations that keep on menacing the community. Planning by itself is a proven process for preparation of emergency plans and hence, planning process forms the key to emergency management. Meticulous planning processes should consist of, in simple terms, pertinent programmes, strategies and arrangements.**

The planning process can hardly be of use if it dosen't reach out to the people or community and their organisations during emergencies ; in a way it is a kind of commitment from which resources, actions, pertformances, costs etc. should flow. An emergency management can scarcely flourish if the planning process is not followed not only as a sequence of steps but also as a distinctive interacting process. Several posers arise in this context particularly in the interactive process of planning : how should we be aware of roles and responsibilities of other agencies ? Can an emergency network be set up with them ? How can this be strengthened ? How can awareness and education be strategised ? What about emergency communication, public warning and forecasts etc. ? The answers to this can, of course, be readily given for any possible crisis through a planning chart. All the answers can be developed in the context of a particular disaster but we need to have a planning process that is fairly generic. We may venture to cite a schematic diagram on the planning process as done by Natural Disaster Organisation, Australia, as follows:

THE PLANNING PROCESS

DETERMINE AUTHORITY TO PLAN
↓
ESTABLISH PLANNING COMMITTEE
↓
CONDUCT HAZARD ANALYSIS ←
↓
SET PLANNING OBJECTIVES
↓
APPLY MANAGEMENT STRUCTURE
↓
DETERMINE RESPONSIBILITIES
↓
ANALYSE RESOURCES
↓
DEVELOP EMERGENCY MANAGEMENT ARRANGEMENTS AND SYSTEMS
↓
DOCUMENT THE PLAN
TEST THE PLAN
REVIEW THE PLAN →

Each step can be further amplified as sub-processes with further delineation. For example, hazard analysis process may be schematically shown as :

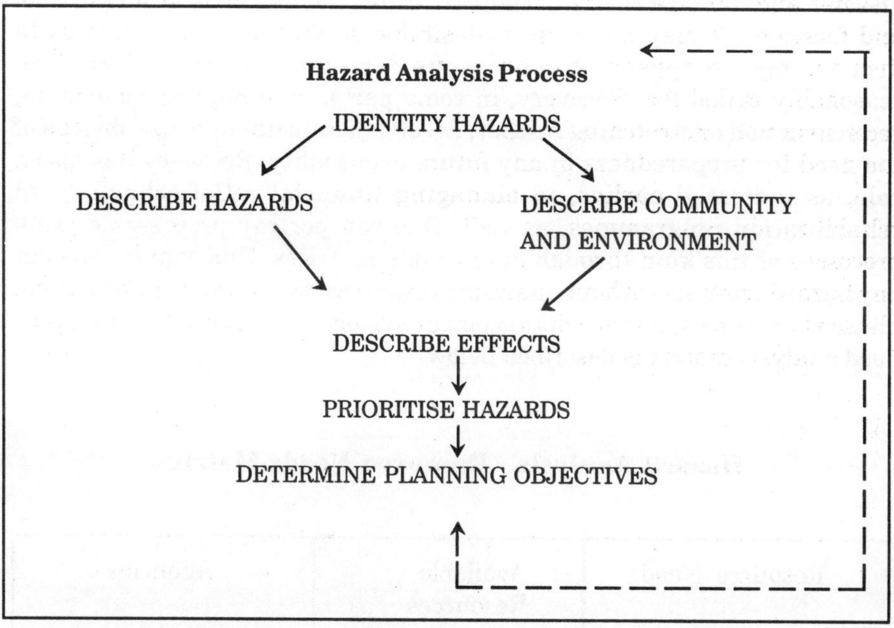

Any kind of process model of crisis management, after the above three 'p's, should go in for four 'r's in general. These four 'r's have some commonalities so that, broadly speaking, we can, for the present, look for two processes, namely, response and recovery.

4. Response : Crisis response should be a single step process for decisive implementation. Response to a crisis depends on the situation-bound leadership, of course, backed up by use of expertise and consultative capabilities. In general, there should be key executive positions (of course, as and when necessary, willed by policy makers, may be political authorities) who should act effectively in the process of crisis decision making. Apparently this may sound as something foisted from the top but the compulsion of the emergency character of situations cannot but allow some sort of centralisation and concentration of authority of the highest possible level. But the assertion of power at that level should necessarily allow transition from the formal level to non-formal leaderships through volunteer organisations, NGO's etc.

Unless there is a studied and conscious endeavour on the part of the top most authority, top to bottom approach as the core factor of planning process may become a futile exercise. Hence, the assumptions and consequences for this phase should be specifically adhered to and monitored, too. Operation system in jeopardy needs to be activated.

5. *Recovery* : It will be too simplistic to treat recovery as the process aiming only at restoring the damage scenario, back to normalcy. That a prior *insitu* knowledge and information, is necessary, goes without saying. It depends largely on the crisis manager to assess, while analysing the disaster and damage thereof, whether restoration per se is at all possible and feasible ; it may not even be desirable so that a fresh proneness to disaster may reappear. A courageous decision making is, therefore, reasonably called for. Recovery, in some parts, may require renovation, reconstruction or stretching further, development without being obvious of the need for preparedness in any future eventuality. Recovery has also a role, as indicated earlier, in managing financial relief schemes and rehabilitation programmes, as well. One can portray processes or sub-processes of this kind through appropriate matrices. One cannot rule out any hazard analysis without analysing resources, for example, information, physical resources, finance management systems etc. A possible resource–need analysis matrix is described below :

Hazard Analysis : Resource Needs Matrix

Resource Need	Available Resources	Agencies
Prevention		
Preparedness		
Response		
Recovery		
Relief		
Rehabilitation		

Remarks

1. Crisis prevention may assume complex dimensions because of vulnerabilities ; hence, the need to attain as much perfection as is possible in respect of prevention schemes and safety measures.

2. Anticipatory and prevention exercises should in no way be oriented so that the capabilities on resilience and response suffer from being developed.

3. Planning should in no way be a victim to adhocism so that a crisis is followed by another one at the conceptual and doable levels.

4. It should be remembered that expertise wherever they may professionally belong to, may also labour under psychological handicaps and particularly behavioural types of crises so that rationality in decision processess may not suffer.

More about Crisis as a Prelude to Crisis Management (CM)

A general impression about Crisis Management (CM) is that CM deals with man-made or human-caused crises. Some tend to distinguish between CM and other forms of management that deal primarily with natural disasters; indeed, emergency and risk management deal primarily with natural disasters. Can we now look into essential basics of CM? Assuming that natural disasters happen as unprecedented and unscheduled phenomena, crises are in no way inevitable and do not need to happen. The purpose here now is to widen the framework of CM so that elements from CM can be drawn upon to grapple with disasters.

Before we proceed further, we need not recall the working definition of a crisis given earlier. We simply assume a crisis to be an event that affects or has a potential to affect the whole of a system. The system may be an organisation or a part of an organisation. If the crisis takes a substantial toll of human lives, properties, financial assets etc. of a community, then we can call it a major crisis. Similar to that of happening for a disaster, a major crisis affects not just the community in the place of occurrence, but has a wider ambit. A major crisis occurs without being aware of what is likely to happen rather without anticipating the consequences of the crisis. Every system, be it an organisation, ought to plan for at least one crisis to happen without being fussy about the types of crises to occur. One might tend to construct models of probabilities of occurrences of such crises and this is an exercise similar to what we are accustomed to risk analysis. If the disaster management is to have the phase for preparedness, that automatically assumes that every crisis is capable of being both the cause and the effect of any other crisis and so, CM is invariably systemic.If CM doesn't adhere to this style of functioning, there is every likelihood that the total management and hence, disaster management will suffer qualitatively.

The best form of CM is preparation for a crisis prior to its occurrence. It is not that a crisis keeps on brewing but there are sufficient indications rather signals for the system to be warned beforehand. This is somewhat a bit different for disasters that happen all on a sudden. There are natural disasters which can in no way be prevented from happening, but crisis is amenable to prevention before it occurs ; of course, the size and complexion of the crisis become points at issue. Mechanisms of early warning systems for crisis and disasters are, in many ways, different. CM is geared to containing damages and so, is disaster management. Performance is a vital element of the management system, whatever it may be, crisis or disaster. Outcomes of performances are equally important. Both crisis and disaster are characterised by common concerns, namely, technology, human behaviour while the components which stand out separately for crisis are organisational structure, culture and hierarchical characteristics. The human element being common to both, manifests largely through responses which influence the individual behaviour as well as the community behaviour. Although in case of natural disasters, rehabilitation is a distinct phase from psychological point of view. Rehabilitation is a necessity of both individually and collectively. Technological factor is, in no way, delinked from human factors, for the simple reason, that technology is run by humans whose mistakes may land the system into a state of crisis. While culture and hierarchy may visibly be well reflected by systems prone to crisis, these two factors may be differently expressed through disaster mechanisms. Many disasters happen affecting communities that have capabilities to contend with disasters ; ethnicity and experiences stand, for affected indigenous people, in good stead and so CM there shows capabilities that need to be improvised in the case of formal organisations.

Civil societies can hardly flourish if they do not reach out to those who are called stakeholders who, in turn, can share the ups and downs of crisis. Stakeholders provide inputs to systems so as to respond to a range of crisis. Although marginal community response systems in case of disasters bring out capabilities after interactions with segments of society outside the community affected and so, there is a shred of resemblance between a crisis– and a disaster-scenario. Can one, then, think of some kind of ethics about a scenario ? In case of disasters, the scenario can in no way, be good, but there can be a good crisis scenario involving the occurrence of a type of crisis ; a good crisis scenario can envisage how the unthinkable can and will occur.

Reviewing further on Contingencies vis a vis Crises

The crisis–reality or state, with reference to contingent typology, has extreme limits, cataclysm and chaos. Disasters coincide with the first part, in general. One can think of several tracks following crises; such crisis paths are obviously time–dependent, either instantaneous or protracted. Likewise, there may be contingent decision pathways that are both analytically plausible and practically verifiable. It may be recalled that there is Hewitt's idea of unness about crises or disasters. We have already mentioned about crisis networking based on case studies on disasters. Indeed, by crisis networking, one seeks to classify "disasters with regard to their decision-making complexity and seeming inability for real warnings to be accepted for what they might become, "the portent of disaster", vide, Rosenthal etal [12]

In the above lines, we have considered how theoretical outfits on crisis and practical disaster realities fit in with each other. Contingent crisis analysis resembles in many ways the totality of hazard identification, hazard assessment and hazard mitigation whose Venn diagram representations may be recalled. (Fig. 5.1).

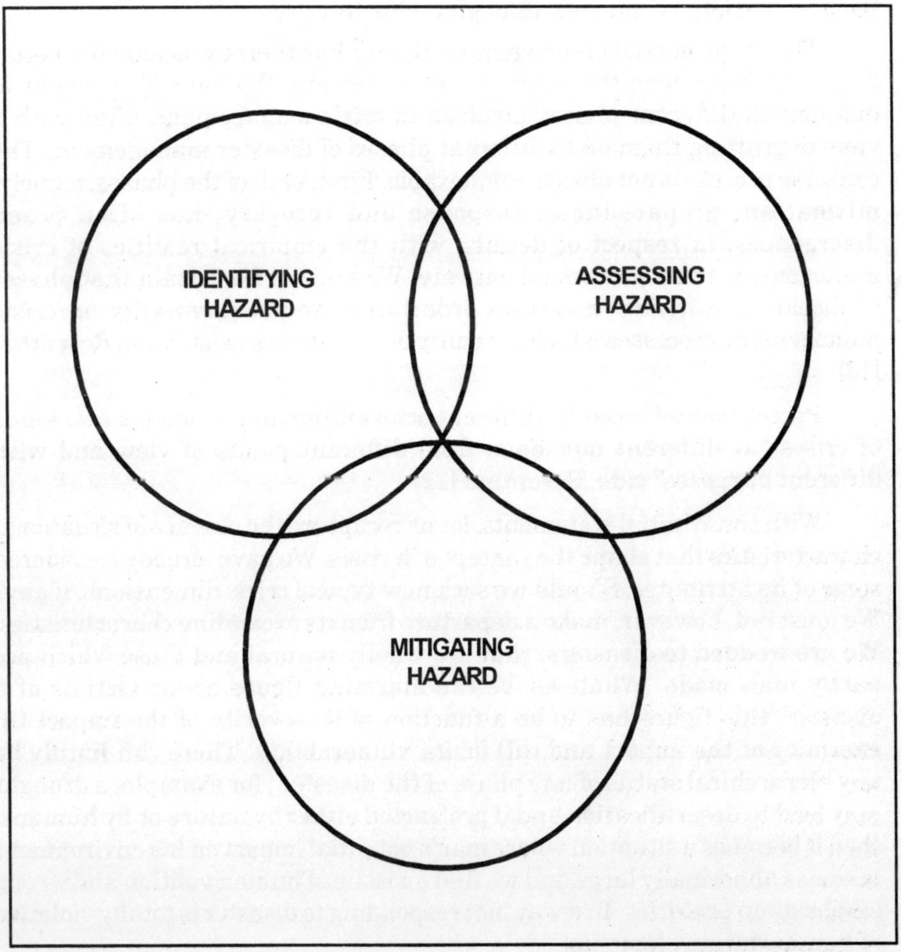

Fig. 5.1

Crisis networking has been shown as an upshot of contingency analysis. We havn't gone deeply into the conceptual and cognitive richness of the contingency-based approach to crisis situations. The factor of predictability continues as an area of vexed concern ; for, if it tends to deteriorate, the technology coupled with environment may lead to something resembling cataclysm. Sometimes, the path to a disaster is protracted. There are, of course, several limitations in contingency analysis, of which two are explicit, namely, analysis of cognitive failure and incompleteness of a crisis path.

Revisiting Crisis Approach with a view to Disaster Management.

On earlier occasions, we have developed a concept of a crisis on the basis of variety of sources that give rise to crises.

The scope of crisis management theory has thereby assumed a better focus. We know who the actors in the crises are. We have also sought to distinguish different phases involved in crisis management, often with a view to grafting them on to different phases of disaster management. The converse process is not always comfortable. First, each of the phases, namely, mitigation, preparedness, response and recovery, has often posed divergences, in respect of details, with the empirical realities of crisis management in its processual entirety. We know it for certain that phases of disasters warrant "a certain orderliness and sequentiality on crisis management processes which in reality do not always exist" vide, Rosenthal [13]

Perceptions of crises by different actors differ and so, lead to new kinds of crises "at different moments, from different points of view and with different purposes" vide, Rosenthal [12]

With these initial statements, let us recapture the essence of situational characteristics that shape the concept of a crisis. We have already considered some of its attributes. Should we seek now typical crisis dimensions, if any? We must not, however, make a departure from its overriding characteristics. We are wedded to disasters, that are wholly natural and those which are partly man made. Whatever be the alarming figure about victims of a disaster, this figure has to be a function of (i) severity of the impact (ii) enormity of the impact and (iii) *insitu* vulnerability. There can hardly be any hierarchical status of any phase of the disaster ; for example, a drought may lead to desertification and if protracted either by nature or by humans, then it becomes a situation where man's potential impact on his environment becomes abnormally large and we find an issue of human volition and wrong emphasis on priorities. In a way, not responding to disaster is totally violative of humanitarian elements.

Redefining crisis becomes then a sort of necessity. The variables about which we talked earlier, continue to exist, of course, with their new dimensions. "The unness provides, as before, crises or disasters". Although included within the ambit of unness, uncertainty has a distinctive role in the definitional concept of a disaster. Apparently, the mundane factors like ambiguity and confusion fall within the purview of uncertainty. Any planned effort or the urge to do so to the contrary is, doubtless, to obtain certainty, without being illusory about it. However surprising it may be, uncertainty and surprise may serve, of course, as important levers for whipping up mobilisation ; in no case, we can afford to dispense with uncertainty. The factor of urgency is a strong concomittant so as to be treated as a distinctive category. In the realm of crisis management, we often talk about unimaginable, unthinkable and often, impossible crisis scenarios. There exists a propensity to neglect crisis even if they seem to be visible as a creeping crisis. We know critical events to come up unexpectedly, despite the best of efforts to avert it. Can this type of crisis be named as a compulsive crisis? So we find the original notion of crisis assuming a multitude of definitional characteristics. It would be in the fitness of things to consider crisis paths based on case studies on disasters which have happened in realities. The development of crisis and responses, too, can be brought to the fore. In regard to crisis management, the behavioural factor cannot be dismissed and more so, the process management involved in prevention, preparation and response. Training and sensitisation, of course, on crisis management should go on unabated.

6

MANAGING EMERGENCIES AND CONTINGENCIES

Any dominant aspect of disaster management cannot shy away from the consideration of emergency involved in disasters, whatever be their nature. Some sort of emergency measures becomes a compelling activity. One cannot delink "emergencies" from disaster management. Hence, the need for considering this here.

Emergency Management

Emergency measures are invariably taken up on the spur of time. When a disaster strikes, there is a trigger often instantaneously to spread out disturbances. The effects may or may not be immediate ; yet, the response has to be there readily forthwith. This is well reflected through activities responding to a disaster. This timely action is, therefore, a kind of response, which can well be described as emergency, so as to provide quick and appropriate signals in the aftermath of a disaster. Response, as we know, has a wider ambit but the emergency begins with the very initial phase of response. However, even if response management can draw upon emergency management, the reverse is not necessarily true because of temporary considerations. Managing emergencies can well afford to have uniqueness and identity of its own. Emergency management cannot but be well structurised ; it cannot be arbitrary. The word "contingency" is also used in the parlance of the word "emergency"; emergencies and contingencies have, indeed, commonalities and discernible shades of differences. Let us, now, proceed with the shape and forms of emergency management.

Like disasters which can strike anywhere and anytime, emergencies are facts of life. Emergency preparedness assumes a matter of great concern. Man even though ravaged by a cyclone or devastated floods, continues to be first line of defence. What does an individual do if the disaster is severe ?

Individuals obviously look for help from others and if the emergency thus created by the disaster is well beyond the capacity of being coped with, they seek for assistance from larger sources. This accounts for setting up of organisations solely to deal with emergencies.

Let us now deal with some specifics, mainly preliminary imperatives for anything to do with managing emergencies. Emergency management, it is to be kept in view, is, naively speaking, the art of dealing with disasters and this sort of management keeps on developing new systems for coping with larger and larger troubles ; indeed, one does not have to wait till a disaster happens so as to teach /learn the fundamentals of emergency management. If one belongs to a local emergency management team, one becomes automatically concerned about preparedness for emergency by the community to which one may belong directly or indirectly. There is no doubt that there is an innate desire to help out in times of need ; we often use the word "empathy" in such situations but that's not enough. If someone is to build up community's emergency management programme, it cannot take off without a good foundation. For, with a well-planned emergency management system, individuals are made aware fairly well in advance to work as a team before, during and after, emergencies ; one doesn't have to be scared about a heavy loss of lives and properties. Planning for emergencies, obviously, assumes a cultural dimension ; it becomes all the more vital in all the phases of an integrated emergency management process. The relationships between the hazards affecting the community and development of emergency operations cannot be understood if there be no prior planning in regard to emergency. Positively speaking, one can augment the plan for emergency operations if there be an effective strategic planning on emergencies.

A value of conducting emergency enterprises is always associated for assessing and improving an emergency plan ; this value needs to be well fathomed only through a systematic emergency planning. If in the initial stage, one finds oneself in a local emergency team, a novice, one has to develop a capacity of acquiring the better grasp of a wider scenario on disasters. This kind of emergency management can exist in any organisation whether in civil defence, medical service, army or any where likely to be initiated by an emergency manager. A volunteer group can also have such tems.

It may thus be taken that a solid foundation on which one can build up an in-depth understanding of the processes forms the broad field of emergency management. Really speaking, each sector will provide for something meant for the community affected. Every disaster situation throws up a challenge and an analysis of the concerned scenarios in terms of real life situations, can bring out, in a manager or the worker, the best that can be provided to the community in danger. Some sort of professionalism, without losing the missionary zeal, has to be imbued from the very beginning; self-learning on the past disasters really empowers the emergency managers or the field workers to have the abilities to reach out, even to those who may not be strictly within jurisdiction of preparedness. Therefore, in sum, the essential elements of acquisition for the management of emergencies are : (a) the need for confronting confidently with impacts of hazards which the community is facing (b) development of enabling mechanisms so that the threatened community can participate in the processes of emergency management and roping in other segments of the society who can help the community respond to the crisis. (c) interaction with participating agencies so as to allow them to put in the best of their skills in order to facilitate the manager or the worker, undertake the enterprise of emergency management as a coherent team, (d) application of the process of emergency management in terms of the outcomes from each of (a), (b), (c) and (d).

It remains yet to dwell on some peaks of emergency management that boil to four essential basics as considered below :

The first is about leadership and influence ; it cannot be visible unless the abilities and styles of functioning are well focussed. An important element, to begin with, is to diagnose any problem in terms of motivation, inter-and intra-relationship in communities' and groups' dynamics ; it ought to end up with remedial efforts. The second is about decision making ; clearly, it is a cue from the first one in that problem solving is indelibly linked with the basics. Sound decisions are called for along with the ability to do so increasingly. The group dynamics, as mentiond earlier, has to be well monitored so as to lead to group decision making. The other two elements adjunct to the problem solving such as identifying and facilitating the processes involved, should be emphasised. Third, the issue of effective communication. Any emergency management can ill afford to miss the analysis of communication issues and so analysing the issues, is the first step. The role of the public, speaking rather public in emergency situations, can hardly be overruled ; therefore, verbal presentations become necessities, to cope with issues arising from any emergency management. Both intra- and inter-communications can rake up issues which need to be resolved through non-verbal means. The liason with the media should also be gone through. Fourth and the last is about financing which is different from the usual support and aid from agencies. It becomes then concerned with non-conventional ways of raising funds so as to meet needs of the programmes. Mobilisation of resources, mainly financial, should be the main strategy in this aspect of emergency management.

As is well known, emergencies occur without any prior intimation ; lives and properties then hang in the balance and hence, decisions are always then found to be something with a difference. One finds in emergency situations shifts in roles and there are discernible changes in regard to rehabilitation, particularly in respect of communication.

Apparently, the policy undergoes a change and leaderships become more concerted and pro-active than what they used to be earlier. New players appear at the centre stage and one finds during the emergency a fairly substantial difference between unorganised or less organised and structured emergency responses. We shall take up policy aspects of emergency management later on but wind up now by impressing on the urgent need for sensitising each one in the whole spectrum of emergency management. It hardly needs any reiteration that every community has to have well planned *insitu* emergency management policies before a disaster occurs.

In the foregoing lines, we have not discussed about possible experiences that can provide a learning curve in how to manage emergencies. Disasters take place with all the horrors bringing in their trail ; even a moderate disaster can leave a community devastated. Therefore, protecting the community becomes a primary imperative of the government and fulfilling this responsibility is largely dependent on the abilities of emergency personnel to respond predictably when a disaster strikes. One should then go in for developing and maintaining a quality level promptness so as to act upon effectively under terribly precious situations. It is always expected that emergency personnel will acquire the quality of readiness and the ability to function effectively. Managing emergencies or participating in such exercises provide a kind of learning process. This necessitates the creation of a structured learning environment which enables emergency personnel to function in a realistic emergency situation. What one can have is the portrayal of an emergency scenario, to begin with, unfolding itself in sequences in keeping with phases of a disaster, that is, pre-during-post. The events related to the scenario, keep on evolving with increasing complexity, threat and unforeseen pressures of one kind or the other. The learning profiles on experiences enable the participants not only to develop emergency policies, plans and procedures but to ensure an effective response. While the scenario culminates in an emergency exercise, the participants do acquire knowledge, awareness, flexibility, leadership and inter-connective skills, often under duress. Any kind of exposure to such scenarios is a sort of rehearsal in terms of real life roles in a realistic emergency situation. Emergency scenario is built up because of its unfolding elements on the basis of community data. There will, obviously, be upcoming emergency exercises and exercise facilities, too. This calls for an integrated emergency management plan addressing to all the segments and players, as mentioned above. Several problems some of which have been mentioned earlier, crop up during the sequences of unfolding of an emergency scenario; these are (a) what kind of command or relationships among them can be, even within disciplines ? (b) individual disciplines are to be co-ordinated for ensuring official handling of any event during an emergency. Moreover, information flow between individual organisations and disciplines should be well taken care of. These are basically issues that should be reckoned for any operation plan during an emergency. One cannot overlook potential shortcomings in emergency plans so that some of them be well taken care of prior to a actual emergency. Any effective emergency management requires teamwork, communications, and working relationships cutting across organisational barriers in their entirety. Unless emergency management is professionalised as a process reflecting ultimately the exact science of managing disasters and emergencies, we can hardly reach a semblance of a benchmark effort in this direction. One has to remember that we don't, honestly speaking, manage emergencies rather we make the community and ourselves too, prepare for them and/or respond to them, which has been well highlighted earlier.

Remark

Having set forth the broad perspectives for professionalising emergency management, we need to consider other adjunct features that will bring to the fore, the integrated character of emergency management without losing its essential ingredients, namely, the human being, which we shall do, beginning, of course ,with contingencies.

Contingencies : Some Preliminaries

It can be contested why contingencies are to be considered after considering crisis. The immediate justification is that both contingency and crisis are generic concepts ; of course, they have mutualities and loosely speaking, overlaps. Whatever be the dictionary meaning, we can have a working definition of contingency to refer to alternative futures including adverse developments. There is a clear-cut domain for contingencies. Contingency analysis and contingency planning are here to stay. If we consider alternative futures, we necessarily reckon the multiplicity of futures rather than the future. It becomes, of necessity, then what these alternative futures are about ; infact, are these identifiable ? There has to be necessarily conceptualisation in respect of contingencies. For example, contingency thinking which leads to contingency management thereby setting at rest any sort of uncertainty and ambiguity of an approach identifying an alternative future, we cannot but take into account some estimates of probabilities. Short term variations and long term trends need to be also taken into account ; there can also be linear or non-linear trajectories to the future. How can contingency analysis and planning be sensitive to such criticalities ? It should be kept in view from the standpoint of contexts and processes that contingencies should have crises as their allies, even at the conceptual level. An alternative future set out by contingency analysis may bring about acute crisis which may figure extraordinarily in materialisation of an hazard to a disaster. Inter-disciplinary textures do obviously come up when we still consider both contingency and crisis analysis. Whatever be the lines of demarcation between crisis and contingency, the conventional risk analysis may not suffice for deciphering the origins of neither the crisis nor the contingency ; both of them require a qualitative analysis because of the emerging complexity of systems involved. Sheer complexity may bring about an increasingly vulnerable character of systems, in particular, their infrastructural elements. Naturally, the world of contingencies, and of crises as well, acquires a greater competetiveness that what it used to be earlier because of globalities involved. Such a global scenario poses an unique challenge for assessing the opportunities brought about in the wake of possible alternative futures and contingencies. For developing countries like ours, massive efforts should be brought to bear on potential assets and resources ; mobilisation on this score is warranted behind the shield of adversities and ordinarily speaking, crises. Isn't it a new but difficult, kind of management of opportunities by thwarting exploitation from new contingencies and crises ? The challenging tenor of the opportunity management will hardly flourish if there be no pro-active players in a competitive world of contingencies and crises ; this requires some sort of adroitly structured restraint of a substantive degree. This kind of conceptual genericity, in regard to contingencies and crises, will provide enabling means to turn future effects of disasters (crises) into opportunities of confronting them. Therefore, a good deal of learning should be in store in regard to contingency analysis and emergency planning and so, worthwhile attempts should be made to build up capacities for equipping the takers with learning modes. It may so happen that the knowledge, from the past, in the learning mode, may have some shortcomings stemming from some disturbing happenings. This may even go to the extent of coming in the way of acquisition of knowledge about risks and contingencies. This obviously lands the policymakers in the arena of ambiguities and uncertainities. But, isn't it a challenge to the domain of organisational and policy learning so as to be sensitive to adversities ? Can we overlook opportunities, trends and singularities ?

While we emphasise on awareness and education in regard to disasters, with the community as the dominant clientele, for contingency management there are learning efforts on a continuing basis, in which contextual and thoughtful knowledge are well pursued rather than taking recourse to ritualistic approaches to decision-making and implementable processes.

Remark

Theorising contingencies in some of their facets, becomes a necessity because of contingent (crisis) management. Let us therefore, explore how to go about applying contingency analysis.

Theorising Contingency vis a vis Contingency Management

We have already talked about the context where contingency becomes a matter of concern, to be acted upon with important skills. We have to realise that contingency-related futuristics can occur and contingency may stem from complex situations. This view seems to have been glossed over as a critical element in devising strategies. We haven't mentioned about what are often described as contingency costs. That is, in simple terms, we pay little attention to requirements of increasingly contingent crisis tasks, but we can hardly afford to miss it, if situations call for innovations. In complex situations, we cannot anticipate neither the wherewithal nor the termination, so that any sort of strategy becomes inevitably sensitive to contingency. We can, then, rarely avoid a highly contingent task.

Complexity, now a days, for any organisation or endeavour, is fast assuming a variety of characteristics, not simply because of the context but because of the realisation that the system concerned does possess self-organising abilities. This becomes all the more explicit because of the participation of human elements through community response. Creativity, then, becomes a continuing outcome. The task of mobilisation of people puts an onus on the leadership. This cannot be realised without co-ordination; more than that, it requires a kind of concurrence or convergence of behaviour to which we may turn later on. Let us remain content with the fact if the community response or even the response of the leadership is to be qualitatively innovative, then limits of concurrence or convergence must be, even if informally, be kept in view.

With the foregoing remarks, let us now proceed towards applying the dynamic principles of contingency thinking to respond to disasters. Any analysis from the standpoint of contingency, for example, in regard to coping with tropical cyclones, we must have the concerned technology and environment around, to be two significant variables, often called contingent system variables. Coming down to more realistic levels, we may have size, structure, complexion, texture, composition etc. as examples. There may also be, in view of complexity, a few more contingency variables, which need to be considered as those preemptively given. Although, seemingly harder concepts, sensitivity has to be reckoned because of varieties of uncertainties. It is true that technology which has already been taken as a contingent variable, affects structure, leads to interactions and exerts some imperatives on individuals, as is well borne out by responses to forecasts and warnings, using technology. This is how, one can conceptualise contingency models. Without going into the outcomes of contingency analysis, which are, of course, in many ways, amenable to deeper study and research, let us see how such models can be of use to practitioners. Any participant in a disaster or crisis situation must seek a match between the situation and the concept about it, and also the complexion of the match itself. It is a truism that the participants, players, analysts and others, in the exercises of disaster management, whatever be the phase of a disaster, have to look back so as to exert on elements of reflexivity among themselves. Practitioners of contingency models should see to it that no sector has an edge over the other or others or no sector can shy away on some pretexts ; for example, the non-technological sector where involved in analysing contingencies shouldn't be allowed to be aloof from what technological element contributes to the contingency analysis. A holistic view, therefore, is earnestly called for ; synergising findings and contingencies in the whole framework, whether in theory or in practice, cannot be lost sight of.

In the pre-disaster phase, preparedness or anticipatory steps, crisis reality needs to be looked into in terms of choices by select few and also contingencies that impinge upon individual activities. The contingencies, there and perhaps along with them, constraints may be strategically chosen which may be mistakenly exercised by individuals concerned. The environment of choice by individuals gets inspired and volunteeristic susceptibility or inclination on the part of individuals of the community may suffer erosion. Therefore, the potentialities of individual behaviour and cognition assume a dominant character.

7

RISK VIS A VIS VULNERABILITY MANAGEMENT : A CRITIQUE

Introductory Remarks

It is no way untrue that risks stem from hazards, natural or otherwise. Therefore, any venture on building nationwise disaster management requires risk management. There are no doubt advances in recent times on risk management techniques and in fact, there are new endeavours on assessment of risks of technologies. It is being increasingly realised that, even for governments, risk management is the key to continuing economic development. Further, in these days of sharing partnerships between the government and industry, the latter should accept risk management as a pre-requisite for doing business. There are many elements which are addressed by risk management systems and importantly, emergency planning and crisis management which involve the government, the community and the industry. It is often remarked that natural disaster management can afford to have a conglomerate group or team that can have, by and large, an win-win opportunity for sectors of the society that can put stress on PMP strategies rather than purely on relief. Risk management by the industry, as a joint venture, can be pro-active in planning, logistics and other resources. An inkling of risk management should have risk communication as a predominant element ; risk communication must necessarily be coupled with the participation of the community so as to facilitate not only for impact prediction and evaluation but also decision.

Characterising Risks

There are, doubtless, commitments in respect to a certain risk, whether for a man of the masses or a professional decision maker. For decision makers, the responsibility is somewhat greater because of being associated with risk and also because of the expectation of the masses that decision makers should be resolute in confronting with adversities.

There is every likelihood that the policy or the decision-maker may land in error with respect to the rationality of the risk management/ assessment process. In the foregoing lines, we haven't considered the basic concepts of risk perception and risk acceptance and these are taken as assumed. There is now an emergency concept of personalising risk which is taken to be an important link between knowing about a hazard and taking self–protective pro-activity. A risk manager has to see whether there are significant linkages between risk perceptions and the ability to accommodate hazard disturbances. There are valuable instances where personalising a risk has paid dividends. Risk perception is generally taken in terms of individuals' expectations about the likelihood and severity of disasters.

A contemporary variant of risk perception lies in the ability or strictly, the frequency of thinking aloud, the information about a hazard and in preparing the affected community for adoption of mitigation and preparedness measures. Thus there is what is called the concept of hazard intrusiveness as a factor explaining preparedness behaviour, as can be seen in Lindell and Prater's (200) [15] findings that intrusiveness significantly predicted hazard adjustments even when risk perception did not and also that it predicted adjustment better than hazard experience or a range of demographic variables.

Risk Perception : Role of Gender

The role of gender risk perception is in many ways critical. Women, in general, bear a disproportionate share of risks and therefore they have completely different views on risk, and ways of encountering them. Their understanding of a risk rather in matters of risk communication, is different for several reasons. There is the overriding issue of social acceptability of risks which varies between different gender and ethnic groups. These points are well borne out by what have happened in recent tsunamis. Of late, there have been some excellent sample studies on gender based differences in risk perception, risk characteristics, risk rankings and risk regulations. Once these comparative issues on risks are reckoned, there is every possibility that risk communication can have strategies incorporating the findings of the above mentioned studies. Answers to questions on unevenness of risk exposure, the imbalances in the control of risk, are likely to facilitate the assessment of present and future risks and policies as well, to management. It will, indeed, be revealing if there be findings on risk perception with a cultural and social bias ; we will be able to study patterns of risk perception. There is an emerging view that findings on risk and culture (conceptualised often with cultural biases or politically abiding ideologies) can well become attributable for patterns of risk perception. This buttresses the view that "individuals perceive a variety of risks in a manner that supports their way of life", Aaron Wildavsey and Karl Dake, vide, [16]

Exploring Relationships : risks and disasters

Having considered some aspects of risk management, somewhat component wise, one may feel inclined to take such a critique to be unrelated to ground realities of disaster management. Therefore, without venturing to go the other way round, somewhat negatively, we like to proceed, in what follows, the close relationships between risks and disasters.

An immediate statement should be that we should explore whether a natural disaster reduction, rather its phases, has their counterparts in risk reduction strategy. Any typology of risk reduction strategy must consist of (i) choice of the site (ie. about change of floodplains, relocations) (ii) reduction of loss either through altering the event or affecting the cause (i.e. technological modification) (iii) pre-empting the loss (ie. through warning system). The fourth category is often sought to be worked out through redistribution of losses (either through pre-disaster insurance in some countries or relief work in the aftermath of a disaster). The essence of these measures can be clearly distilled into what may be designated as disaster (or risk) mitigation measures.

We have mentioned choice which presumes adjustments, on the basis of what has been described earlier as perception of a risk ; obviously it does resemble what is called disaster preparedness. Evacuation is also, as we know, a part of mitigation and we shall deal with this in a later chapter but we should mention now that evacuations are basically pre-disaster protective responses, resulting from some form of risk perception.

Revisiting Risk Management

Coming back to risk management, we should keep in view that in course of managing risk, there is every likelihood of being confronted with the question of choices about risks and hazards. Several questions arise such as : how does the community naively make a balance between risks and hazards, keeping in view the ultimate benefits the community may possibly accrue ? In what forms do these become acceptable, separately, taking together or otherwise ? Managerially speaking, how are the concerned persons distinguished between these two ? Obviously the latter group will plunge into the usual path of risk assessment, predominantly as a tool. While risk assessment can go in for exploring quantifiable risks, they are likely to get stuck up when they get to peoples' problems, which essentially is a social process. Social contexts should not be looked upon as an extra-risk assessment affair rather as something that provides substantial and significant inputs to the totality of risk management. Risk management is thus led to encounter some sort of uncertainties or hunch because of science on uncertainties, has to reconcile with its limits. This state should not be taken as un–or anti-scientific because of the fact that ideas on limits to prediction and predictability *per se* have come to stay. Indeed, in crisis management, we need to reckon *cataclysm* and *chaos* as extremes of a crisis and both these notions are in a way offshoots of looking from the standpoint of science, as mentioned above. In terms of brasstacks, in order to reduce uncertainty, a management strategy seeks essentially some form of technological outputs so that the implementation of redundant safety systems are ensured. Further, as there is no such thing as zero risk, the relevant risk management approach should aim at optimality, or strictly speaking, minimality.

In fact, one of the purposes of risk and hazard assessment ought to be to minimize exceptionable elements for the hazard management personnel, whatever be the uncertainties. The consideration of optimality requires, in technical terms, constraints particularly on the limitations of impacts of natural disasters. The global implications of any disaster can hardly be set aside ; the recent Sumatra tsunami striking the shores of India, Sri Lanka and others bear testimony to this. The approaches to management have to differ before and after the disastrous event. We shall later on deal with the evolution of disaster environments.

That risk assessment forms the bulk of risk management is well taken care of by using risk assessment towards decision making about the hazards which in turn bring about risk. What should then be a necessary perspective about the magnitude of the risk ? May be, some sort of comparison will be of help ; but contrasting risks will be more useful, in that, such an approach can give an indication of ways, perhaps alternatives to each other, they are treated in the community. There is, of course, the conventional method of risk being averaged over time but shouldn't we then miss peak-risk ? May be, we may use statistical figures to comprehend the nature of the risk and to express it as well but the persistent concern is the cost involved in reducing risk. There are accounts of endeavours in this direction, the people often being prevailed upon by the maxim "an ounce protection is better than a pound of cure".

Tsunami designs have been concerns with the erstwhile Soviet scientists. They were reportedly as mentioned in some of their presentations, enamourd of some methodological procedures ; these are designed so as to provide "reasonable upper bound estimates" vide Anderson [17]. Therefore, by design, the point estimate of the risk appears at the topmost point of the real risk distribution curve. People still raise the question about the why of usefulness of risk assessment. There can be several answers to this. It is true that risk estimates provide tools and necessitate data for interpretation. Risk management used to be a monopoly of professionals who could hardly care for public needs and concerns. The spectrum of professionals has widened over the years so as to include scientists, lawyers, regulators besides the professional managers and so, there is a greater need to do with risk communication than what it used to be earlier.

Vulnerability Assessment : Why and What ?

Any observation on any aspect of risk management can hardly be dissociated from what is often described as vulnerability. assessment ; but why is there a concern for vulnerability ? Vulnerability by itself has acquired a global stature because of a variety of issues and programmes, particularly those relating to climatic change. No one can afford to be evasive about variegated susceptibility to global changes. Vulnerability reduction becomes then incontestably a point at issue. How do we conceptualise vulnerablity ? There is, indeed, the use of the term vulnerability in a variety of intellectual traditions. It would not be irrelevant to recall what Aristotle and Comte spoke about vulnerability. To them, "vulnerability is a term of such broad use as to be almost useless for careful description at the present, except as a rhetorical indicator of areas of greatest concern", vide Timmerman[18]. In terms of an hazard, one can define vulnerability, as proposed by Timmerman, as the degree to which a system may react adversely to the concurrence of hazardous event. Aren't these commensurate with the recent leading concerns of the World Conference of Disaster Reduction (WCDR) held at Kobe, Japan, from January 18 to 22, 2005 ? One should add that resilience is also what WCDR mentioned pointedly and here again Timmerman [18] links vulnerability to resilience, "to the measure of a system's capacity to absorb and recover from the occurrence of a hazardous event. Apart from being linked with the concept of resilience, there are other concepts, where the concept of vulnerability finds resonance such as : Susceptibility , adaptability, fragility and nonetheless, risk", vide Liverman[19]. Our concern here, is to find ways of determining, who is most vulnerable to natural disasters and why determination of vulnerability, of course, can be found in many of the deliberations on global change and more so, when Earth System Science (ESS) is considered to be the pre-dominant component of disaster science and technology. Community at risk is found to populate areas of high hazard magnitude or frequencies, severe rainfall deficit, (drought) , floodplains, coasts with severe storms etc. A number of regions and countries prone to such disasters are identified as critical zones, eg. Bangladesh, Sahen etc. whoever may be the players in bringing about vulnerability. Economic and social dimensions of vulnerability should necessarily be considered besides the sensitivity of a physical system, while assessing the impacts of disasters. In these days of emerging technologies, one cannot lose sight of the relationships between technology and vulnerability ; of course, technology is resorted to for reducing vulnerability which is the otherway of mitigating some slow onset disasters like droughts. One may refer to dams and other technological infrastructures as efforts to prevent flooding, storm surges etc. In the Indian context particularly in mid–eighties, the years of severe droughts, one could witness an overdependence on irrigation which essentially means greater vulnerability when water storage is at stake; thus technological constructions carry the potential for disasters failures, thereby encountering a high vulnerability to devastating floods. One cannot rule out that a social divide can be brought about by an increasingly fragile social inequality and vulnerability. Who can deny the social marginalisation of small farmers and other disabled groups with new-fangled technology ?

Let us, therefore, sum up, in generic terms, identifying potential measures of vulnerability. We have already emphasised the need for an analytical approach to vulnerability studies, but it is equally important to have a more synthetic approach to take on vulnerabilities. It is not always true presumably contrary to what have been stated earlier that socially vulnerable people may not stay in vulnerable places because of compelling economic needs ; it is also true that fairly well off people live in relatively fragile environments ; so one finds often a mix up of two types of vulnerability that may stand further impoverished in regard to risks involved. Mathematically speaking, can we then seek a mapping, in a geographical space of vulnerability, to a social space devoid of inequality and inequity? Of course, we should have the apt definitions of both the spaces. If we are to be more analytical in our approach, we should have necessarily a specific scale of analysis or what is called a reference unit. Considering difference from one scale to another, one can conceive of cascades of vulnerability but there ought to be different factors determining vulnerability in different cases. One can identify variables and conditions in determining vulnerability in general terms but we shall confine our remarks in the light of natural hazards. There do not exist many worthwhile investigations in this direction but what stand out to be immensely valuable is that there should be objectives to obtain a better understanding of vulnerability, too ; at least, some natural disasters in terms of terrain access to technology and environmental susceptibility. For example, in the case of a slow onset natural disaster, say a drought, the pattern and severity of loss for this disaster may be worked out, one should also determine the relationships between drought loss and the physical pattern of timing, we should find out in this process, variables to represent the framework of vulnerability to the natural disasters and then account for pattern of losses. The significant differences in losses between different forms of disasters will eventually emerge.

Many questions about vulnerability assessment seem to be unresolved. As a part of this sort of exercise, we need to know how to deal with a multiplicity of vulnerabilities to a multiplicity of natural disasters. Unless we know the pattern and degree of vulnerability with respect to current natural disasters, we cannot use it as indicators of future vulnerability and in consequence, the communication about vulnerability to scientists, managers, activists, communities and the policy makers will not be fully and properly apprised.

In sum, we can hardly be aware of social impacts of these events unless we decide to work for understanding, recording, and communicating the complexion of vulnerability and this forms, in a way, the crux of vulnerability vis a vis risk management. The risk has, therefore, at its core, often overlapping with the vulnerability. Let us now move on to other aspects of risk management, may be a bit distant from vulnerability management.

Some other relevant aspects of risk management

As things stand, now, in respect of calamities, policies ought to be formulated and decided in principle at the national level. Risk management being a significant part of the overall disaster management, one has to see that steps are taken to reduce the influence of political pressures such as organised interest groups lobbying on risk assessment/ risk management. Risk management is taken as a process of valuing and choosing between decision options in scenarios where individuals suffer from natural disaster exposures. What should then be the conceptual construct underlying risk analysis at the central level ?

Risk management techniques however, codified they may be, a sort of reflecting is sore dichotomy, the advantage of which may be taken by pressure groups, as mentioned above. What is desirable at such junctures is that risk assessment should be separated from the wider context.of risk management. This principle leads to the fact that scientific decisions of risk assessment should in no way be dependent on the political decisions of risk management. There is the other side of this point of view, as per the logic of decision theory ; risk assessment has to have choices conditioned by the cost and benefit accrued to the people who may be affected by the decision and so, there may arise from such risk assessment decisions, blunders on the part of risk management, from the point view of general public. One has, therefore, to strike a balance, so that the public do not suffer when risk assessment is separated from risk management. Isn't then, from the point of view of consumers, an optimal risk management strategy is to be sought ? Rigorously speaking, an agency at the central level should develop a risk-neutral model keeping in view consumer's perception, attitude and acceptance of risk. Obviously this is a regulatory task that ought to be foreseen by the central agency being envisaged now a days in the Indian setting.

Coming to brasstacks, if a risk is to be viewed at par with realities of a natural disaster, let us cite what the newsletter *'At Risk'* published under the aegis of IDNDR, UK., has kept on reporting on aspects of risk genre in different parts of the world. UNDRO News and the bulletin STOP DISASTERS used to do this throughout the period of IDNDR. The realities about risk are still brought out vigorously by various leading organisations in different parts of the globe. Like disaster reduction, there ought to be steps for reducing risk under definite responsibilities. Assessing risk related factors has occupied the attention of many responsible bodies and the upshot is more than a mundane survey. One, often, does designate this as a disaster audit. Such an exercise is essentially about disaster preparedness, mitigation and response, vide Sandersonetal [20]. It shows an analysis of strengths and weaknesses of all those involved. Such perspectives provide many organisations with opportunities on devising participatory networking which should be kept on evolving in the context of new patterns of risk and vulnerability. Some years ago, Mr. David Sanderson, it may be mentioned, associated with Oxford Centre for Disaster Studies, participated in a DFID Project in India for identifying risks and reducing measures to prevent them. The project had the "action planning" component that involves all key players, mainly, communities, government authorities and non-governmental organisations and this was precisely about reducing risk in two cities, Delhi and Ahmodabad, in India.

Technically speaking, one can relate risk to safety, time,cost or any viable parameter. The case of a disaster affecting coasts, for the safety of infrastructural elements, requires risk analysis and in case of the recent tsunami hitting the Kalapakkam Nuclear outfit, there still seems to be an element of fuzziness on the safety of operations in such an offshore installation. People, often, talk about accommodation quarters. It ought to be expected that there must have been risk analysis in terms of fatalities whatever be the figures the media say. That the people would be at risk, was well envisaged but still an ethical question continues : can we minimize the risk for individuals ? Can we minimize the overall potential loss of life during the formulation of the project ? It is true that risk managers must have reckoned the diversity of risks around the project and how to prevail on the technical people so that the risk of failures could be well reduced. What often goes unnoticed, is the non-availability of reliable data ; may be in the larger public interest, some confidentiality is often imposed but how would one then design such projects in future, taking into account the overriding factor of safety margin ? Doesn't it limit the capability of a risk assessor and more so, in view of the complexity of real situation ? The complexity as a concept is fast assuming many dimensions. An intuitive risk assessment will not suffice ; what is desirable is to develop a multifaceted study on risks in their diverse forms. A risk manager can ill afford to be swayed by media reports ; of course, media do not have the monopoly in spreading information ; indeed, all of us have, in one way or the other, vested interests in exaggerating aspects on disasters. Perceptions of risk may often be dubious, and they may often run counter to decision-making processes ; particularly, those in terms of ethics and values. Public perception of risks may, however, be bank upon by establishment. How can we then anticipate the evolving nature of risks ? It is apparently a question of prediction or at least an estimate of loss. In the case of a tropical cyclone, we visualise risk through swirling 'vortices' and the track of the 'eye'. In terms of Physics, we can say that physical processes inclined are chaotic. We cannot say anything about the situation becoming more complex and more chaotic but steps can be taken for protection from evolving risks.

Risk assessment has to have identification and evaluation of risks as dominant components. Risk managers have to develop their own norms particularly to distinguish between *de manifestis* risk and *de mininis* risk.

8

MANAGING DISASTER ENVIRONMENTS : INPUTS FROM KNOWLEDGE MANAGEMENT

Introduction

Disasters, whatever be their nature, keep on recurring without losing their identity and uniquenesses. Each of them generates problems and issues, impacting, in a variety of ways, the environments where they occur. The environments are in some ways unique but they do reflect recurring problems. It has almost become a part of the common parlance, to describe environments in holistic terms and recently, in terms of sustainability. These are, no doubt, significantly important but in the light of disaster environments and more so, for their managerial exercises, one should identify the overriding characteristics of disaster environments. An initial attempt in this direction may reveal interdependence or interconnectivity, establishment or organisation and efficiency or excellence as key constituents of a disaster emvironment. For purposes of management, we need to look for the dynamics of such environments. Knowledge specificity becomes a compelling necessity and so, the need for knowledge management.Knowledge management has to have a capacity to convert learning into participatory activities. What we need to explore how in each disaster, emerging during its period, characteristics in regard to its phases. In each phase pre-during, or post-disaster, how the acquisition processes in disaster are facilitated individually and collectively. Can these be optimised ? In this chapter, we deal with some of these aspects pertaining to the consideration of environments, prior to, striking and aftermath of disasters, drawing upon some emerging thought-processes on knowledge in management. We do this in only one section, apparently briefly, but studiedly so, in that the essence of knowledge management may be culled and coupled apparently with relevances.

Rudiments and Relevance of Knowledge Management (KM).

In the context of management, one can readily join the fray in terms of what is often being bandied about as corporate knowledge. It looks that it is narrowly market-driven and production-oriented but the area of natural disaster environment does not consist of only bearings of such a type. No one can deny now-a-days the role of corporate sector in disaster mitigation, relief, renovation etc. The words 'joint sector' have come into vogue in respect of management of disasters. It is equally true that the collection of knowledge by and within an organisation ranges from archives, historical data about disasters and even from, call centres. How to use this knowledge on disasters? How to disseminate it to people who need them ? Organisations have to be well equipped with increasing dominance of knowledge. In order to provide responses, paricularly, to the needs of knowledge on the part of the community, there are several available methods; the simplest, although technical, is the e-mail and that is one way of measuring knowledge which is done through a sharing of knowledge. But for whom ? Are we to assume that there exists knowledge hoarding ? In terms of natural disasters, these terminologies make senses in that the preparedness phase has to be pro-actively geared. In a warning system, the measures that matter most are those which need to be put out to those, who stay at the bottomline. Herein, knowledge management faces a challenge, how to deliver the relevant stuff and how to measure them qualitatively in terms of community response which provides an opportunity for discovering new niches that couldn't be neither foreseen nor did they exist earlier. The financial implicatioms add not merely to the complexity of the process but also the complexion of investment, considering risks and vulnerabilities. Hence we can now, possibly, recognise few significant criteria for knowledge management to be a success: (i) communication, both inter-and-intra, (ii) commitment to knowledge, (iii)participatory network, (iv) innovative community culture, in hours of peril, (v) use of cutting edges of science and technologies. Some of these tenets when applied through a community, ought to yield excellence in cross-fertilization of backgrounds, experiences and innovative enterprises.

Any knowledge-driven activity involving some of the elements mentioned above, should seek to build up the capacity not only to carry on the heritage of knowledge but to create new forms of knowledge. Disaster management has thus strong overlaps with knowledge management. It should not be construed that a knowledge-driven society will take care of disaster affected community if its constituents are ritualislistically grafted on to activities in different phases of a disaster. It is simply because of the human element that the roadmap of disaster management using knowledge management elements, should be distinguishably different. Can we then think of the participation of the community endowed with abilities to create knowledge in a disaster-affected environment ? This group has to go in for "creation of a knowledge intensive environment, along with processes to efficiently create, share and protect the knowledge" on hazards and disaster vide Kalam [21].

The knowledge on disasters essentially consists of happennings. We hear about *the great deluge* as something not only ancient worldwide but described as inevitable. Haven't we heard ancedotes of events that happened some four thousand years ago in Mesopotemia, about devastating floods ? Who doesn't know even now about these disasters through myths and legends? Floods continue to be frequent and often intense, as well. It often looks that even if we describe such events as "natural disasters", can we really call such disasters to be natural ? Isn't the aftermath of such events in all their terrible phases, unnatural? The disaster environment is brought to the fore because of emergence of shortcomings in regard to planning and management.

The services being catered to during each of the relevant phases are in many ways conditioned by the relevant environment. Risk is invariably linked with a rapid onset or a slow onset, type. Environmental risk, as we know, is of a different genre. Hence, the coupling of the two, resulting in the risk owing to disaster environment becomes all the more complex. This calls for tackling the scenario as a sort of emergency or contingency. Environmental planning and management which have come to stay, assume new dimensions in the context of disasters. The compelling services to be catered to the affected community, thus, need to be environmentally sensitive. Once we bring in environmentally conditioned services, we can hardly afford to miss the domineering role of sustainability, which, in turn, requires a cohesive integration of ecological, socio-economic and organisational elements, inextricably linked with allied risks and vulnerabilities. There do occur many providers of services during each phase of a disaster and there also occur threats to environments which assume apparently conspicuous characteristics on account of competitive aids, reliefs etc. A disaster-based eco-system is sure to emerge and therefore, there should be an integrative framework on delivering ecologically oriented mitigation and response services.

In every phase of disaster management, there are problems and issues that need to be managed and if not done well, they come in for one kind of indictment or the other ; for example, the one about relief and recovery. Strictly speaking, whatever be the upshots visible to the public, in general, it often looks that there is hardly any indepth analysis of performance management in each of the phases. Although there has been no dearth of idioms and acronyms for performance management, there has scarcely been any profound conceptualisation on designs and methodologies on this score. Let us seek to venture this briefly. The credibility as well as accountability are often questioned on performance management. Information, as is known, may mislead or often misdirect disaster managers in their performance management activities. How are we to measure programme outcomes besides impacts of a disaster ? Obviously we are in dire need of relevant data. It looks, through empirical analysis that use of adminstrative data in performance management is least likely to provide fairly accurate estimates of the exact figures of impact. But the data may arguably generate useful information for emergency managers ; obviously, because of the fact that one can articulate disaster policy levers improve organizationally or even phase wise performance. There exists a variety of tools of performance management and public expectations in the light of environments being struck by disasters. Like the risk assessment and risk management, one can think of performance appraisal and performance management. One can talk about performance accountability in order to cash in community response, preferably through decentralisation. For every emergency manager, an important question is that he should address to himself is : how can a outcome based system be more effective than conventional approaches ? There may be several posers of this kind : can we think of quantitative evaluation so as to influence in positive terms, programme outcomes ? Does it imply that information obtained apprise emergency managers at multiple levels ? If, on the other hand, there be imperfect information and apolitical complexities in outcome-based system, can there be providers of effective disaster-management, planning and control for improving outcomes ? On programme outcome and impact., one may develop a model, where measures involved are variables relating to sites, control element, for regional and economic conditions at the sites, average outcomes across the site and over the years, on the disaster etc. may be taken into consideration. One may even allow the effects of individual or community characteristics varying across sites and years; for example, the behaviour pattern and responses of the indigenous populations in Andaman & Nicobars, to be accomodated in exercises on modelling ; there may be some modelling in similar endeavours.

Awareness and education continue to be integral parts of mitigational processes ; this may often pose like softer versions. But because of pressing human considerations on the emerging knowledge management (KM), what we require is disaster-knowledge management from human resource (HR) perspectives. HR's continue to evolve with new tapestries so that they provide from time to time, new visions for future directions for HR's. HR activities form the forte of global knowledge revolution and knowledgeable organisations, too. One can go in for, with naivety, grafting elements of knowledge in each of the disaster phases but the more valuable activity should be how to craft in knowledge using the essence of HRs. Peter Drucker, the celebrity on knowledge, coined the word *knowledge worker* ; in fact, in his seminal article entitled "Managing Knowledge means Managing Oneself" he tells about not merely evolution of KM but also about rapidly growing numbers of people with choices and hence, he emphasises on the role of the individuals. It is true that people affected by disasters are distressingly incapacitated and hence, the need of a perspective of a different sort of management by objectives and self control, as per Drucker. The context of the disaster environment, in its wider sense, is not to be belittled ; on the other hand, it has to reckon the growing complexity of the characteristics of the work in each of the phases of a disaster with a view to making decisions in the performance of the work – this bears an obvious analogy with performance management treated earlier. A *knowledge worker* under such circumstances has to transcend his role into that of a *knowledge crafter*.

We know of forecasts and warning systems, as precursors of disaster knowledge. How can HR become a forerunner and leader of knowledge crafting in warning agencies ? One has to keep in view that the warning activity has to have a dominantly systemic view of human capital that HR has. The serious setback that a governmental agency is likely to suffer is that HR there don't have compelling financial incentives to create new human capital models, principles or styles of functioning as fast as in the so called non-governmental set up. In situations experiencing the aftermath of disasters, one has to admit that HRs have to traverse unscheduled paths, and do work at a dangerous pace. Whatever be the complexion of HRs, the purpose is to thwart possible future shocks and this nexus recalls what the great visionary Alvin Toeffler [22] said in 'Future Shock' way back in early seventies and the relevant excerpt is given: "To survive, to avert what we have termed future shock, the individual must become infinitely more adaptable and capable than ever before. He must search out totally new ways to anchor himself . . . . Before he can do that, however, he must understand in greater detail how the effects of acceleration penetrate his personal life, creep into his behaviour and after the quality of existence".

What is missing in counter disaster activities is networking among the people. Knowledge relationships are shareable elements among the people forming such networks. There are many immediate and long term benefits in this way of interactions. If the disaster-environment is to be run on knowledge based principles, one has to make use of people's network that would not only complement existing structures but also facilitate knowledge sharing structure, learning and change. Thus people's network will provide the community response with a vital element of disaster mitigation ; this is how groups of people informally become attached to each other by sharing knowledge and a passion for developing enterprises on a collective basis. Can't we call them "communities of practice" ? One may look forward to such networks to evolve into a centre, replicable in many ways elsewhere. Such a centre can provide contacts within and outside the community, promote interactions on field experiences, prepare a disaster-event calender for members of the community and others well and can work on developing a glossary on leading terms on natural disasters. All these activities, taken in their entirety, enrich the environment which is expectedly dismal in the wake of a disaster. A long term pay off in such enterprises should be to achieve convergence of innovative in HR processes, practices and new technologies so that the gaps, between use of warning systems and acceptance of the risks associated with disasters, can be removed.

It may sound somewhat irksome to policy makers, in particular, about the need of empowering the community with knowledge ; strictly speaking, it is the enterprise of disseminating knowledge to right people through right means. Indeed, one should be cautious about middlemen in this process of disseminating knowledge, keeping in view, the term *disintermediation* which obviously means the act of dispensing with people who seldom empower the community with appropriate needs during a disaster. The great Aristotle observed "action without knowledge is faulty. Knowledge without wisdom is perilous". One has to look at knowledge management from this point of view. Of various slants to notions of knowledge, two types are intrinsically relevant for a knowledge about disaster. Even though it may appear theoretical, a disaster based knowledge has *tacit* characteristics in the sense that there is a notion of disasters in the minds of the individual and the community ; the second one is the *explicit* knowledge about disasters which is the process to extract ideas from people and to put on records distilling the essence of their ideas and experiences, too. Therefore, all that is required in disaster environment is the expeditious need to transfer *tacit* knowledge into *explicit* knowledge. There is, of course, something implicit in these definitional exercises and it lies in the process of disseminating the information collectively held by the community ; other implicit elements are, of course, identification and organisation. Shouldn't we reckon now the distinction between knowledge and information ? What are the distinguishing properties, if any ? Certainly, one should have the measure how an individual can afford to be at ease with the theme of the disaster so that one can not only understand the shades of distinction but one can use it for adaptations. The second criterion should be that the information is necessarily implementable. Doesn't this conform to Aristotle's observations as mentioned earlier ? It is futile to treat KM purely as a piece of technology, although KM does make use of technology. As the recent tsunami has revealed, we need to either buy or build a warning system in the Indian ocean. We can't say that we are going to procure a KM system. What is to be driven home is that agencies should make use of KM disciplines in close alliance with relevant technologies and then, disseminate the same to the community.

The speed in a rapidly changing environment of disasters, is, doubtless, a byword ; the environment demands that agencies should respond quickly. One has to identify the stimuli even if there be a disastraously worse environment. One cannot stay away from reacting to changing environments, even if the necessary information is not available. The recent tsunami has shown the complexity of tsunami waves of galloping returns within a definite duration. The archival of the present tsunami records will provide information for dealing with future shocks of this kind. A close ally to such processes is the flow of technology in the coming years. If it is possible to build up a document as an information system structured for consumption of the community, the criteria of KM, as in above, will be obviously fulfilled. The better the available information system is, the more easily an individual can make use of it which in a way is a kind of knowledgeable action or decision. We can speak of this as an enabling technology for uses of KM.

This is also referred to, in management parlance, as document management, which has, obviously, relevance in the context of disaster but one can ask : how long will a document serve the purpose usefully ? Should it have a life-cycle ? Does the cycle have phases ? How to identify and manage the document reflecting a good KM system ? Any worthwhile KM system can hardly be complete if the internet and in particular The World Wide Wave (WWW) are used as pervasively as is possible. In any disaster affected area WWW with its concommitant technology can take care of interconnectivity and interoperability that are well supported by infrastructures that become visible during the recovery phase. WWW is known to be an important dissemination vehicle and the more we expect it in that direction, the better will be the processing of response systems. While we work for three r's, namely, relief, recovery and response, we should see that appropriate agencies and organisations control the how, where and when, these three r's are executed. As a matter of fact, there appear to be environments corresponding to these three sub-phases. In such a case one may think in terms of the flow of work through these environments with another set of sub-three r's namely routes, roles and rules which being so obvious that these need not be dwelt upon. Yet the work flow through them requires some kind of motion of the associated environmental sub-phases. In this indepth analysis, the question that arises is which agency or the community becomes a predictor of who would be the takers of the system. The recent tsunami shows an extraordinarily motley group of takers with virtually no correlation between a potential taker and his or her tendency to make use of warning services. Therefore, with all the deliberations for an warning system in the Indian ocean like the one at Hawaii, a poser remains as to which agency could effectively predict the likely takers and when KM has to make a dent on this kind of predictive utilisation analysis that may help in strategisation and implementation. KM can hardly do away with a well defined and well structured planning. Typically that plan has to reflect besides genericity, analysis, design disasters (for example tsunami designs) risks and vulnerabilities and finally development so that the work flow in each of the phases of a disaster and corresponding roadmap to success is well ensured, whatever be the limitations of technologies and human resources, as well.

9

MANAGING SLOW ONSET NATURAL DISASTERS

Introductory Words

The previous chapters are mostly confined to management of a few variants of natural disasters. Although strategies of management dwelt upon earlier show a fairly generic character, there are striking differences among some of them, primarily because of uniqueness associated with each disaster. One chapter has been devoted to management of crisis, where we have seen how "creeping crises" as well as "manifest crises" are issues in the affairs of managing them. So managerial endeavours ought to have variations in regard to disasters concerned. One does not have to have a distinctive typology on management of natural disasters but because of intrinsicalities of each of the disasters, the managerial flavour ought to differ. Human element is irresistibly brought to play. Obviously the rapid onset disasters is admittedly beyond the prediction of mankind ; the management on this score has to be limited. On the other hand, the slow onset disasters like floods, and droughts etc. can well be taken care of because of their developing stage by stage characteristics or in brief, their creeping characters. There are, as is well known, involvements of humans in not merely as causative elements but involvement in worsening the disaster scenarios, whatever be the sector of the people. In a way, these fall within the fold of managing disaster environment which we have considered, generally, in the previous chapter. There exist some peculiarities of slow onset disasters which need to be dealt with in depth. This forms the purpose of this chapter and for reasons of not being too unwieldy, the treatment here is confined to flood and drought management.

Floodplain Management

One could have captioned this article as 'flood mamagement' but this might take away the more vital aspect, namely, the management of floodplains. Since the days of antiquity, floods continue to be mythically associated with mankind.

Who has not seen the people in the floodplains rushing to high grounds while anxiously listening to weather reports ? Even now, who doesn't know, how, roads and bridges are washed out ? Don't we move with relief and aid when homes, farms etc. are destroyed, often submerged by the rush of water inland ? Don't we read on the faces of the victims who seek refuge in makeshift shelters, schools, clubs etc. and telling us, their dismal happenings of their relatives or close persons disappearing under the waves ? Aren't these occasions for solidarity of the community and sacrifice as well ? True,

Fig. 9.1: METEOSAT imagery showing the typical cloud systems associated with continuous rain system (Courtesy : ISRO)

we have built up massive dams that are designed to make the people live and farm on the lands to which water belongs but do dams wipe out river valleys, turning them into temporary lakes? We know of evidences of structures aggravating floods. Don't we often destroy wetlands, the natural storage of waters ? Despite all these, we keep on, year after year, preventing floods and the phase of recovery of a disaster is kept on vehemently. No country in the world is in a way immune from subsidising disaster or to put it cynically, promote them more than the other entities. There are staggering stories of the human toll, whatever have been the advances of weather forecasting and communication. Given this distressing portrayal of floods, what can flood management do ? How do policy makers act upon? Does this management have an emergency component ? How should we go about ? Is it the conventional "flood-and-rebuilt cycle" ? Such immediate efforts can hardly be dispensed with even now, atleast to begin with. There are various accounts of inundations, destruction etc. because of floods. It is often said that since we cannot prevent floods, floods are going to happen.

The amount of rain falling on the field is often attributed to human activities. Indeed, raindrop falling anywhere has its path not just determined by the weather or the topography of the land. Agicultural tilling and drainage, deforestation, suburban development, tourist spots etc. by private landowners are all the human actions that the single raindrop comes across in its path. Don't we know the killing of fish species because of run off flooded cornfield? Isn't it that fertilisers wash down river facilitates the growth of algae that takes the edge of water of oxygen ? We can multiply instances of human activity on the amount of rain falling on the land. The recent furore on El-Nino on weather system may contribute to increase of precipitation and *in situ* storm and so flashflood. We have already mentioned about how wetland destruction worsens flooding even on a local scale. Can wetlands prevent the flood peaks which form the cause of damage ? So basically it is the assault on the eco-system by the people that aggravates floods. Engineers keep on harping on nonstructural approach with no discernible changes in policies and actions.

Endeavour on floodplains acquires a truly management flair when decision-making processes are to be undertaken about floodplains. This has all the more become essential because of resources associated with floodplains. Of all the sources, water resources are being accorded relatively higher priority, the reasons of which are not far to seek. Water scarcity and its depletion in years to come, form a continuing global concern. There exists now a torrent of literature, deliberations, activities and programmes in different levels how water resources can be provided a lasting character that can be put to human use. It seldom matters where the water is in the stream or other waterbody or it is diverted and used elsewhere. River linkages have rightly become an issue of great interest. We know about use of surface water for irrigation, for industrial and municipal purposes and even, for production of energy. Obviously, such uses have an impact on the flow or level of water, may be for a shortwhile but undeniably there is a degradation of its quantity

– these are what keep on happening offstream. On the other hand, we have instream uses of water, such as navigation, fisheries, wildlife habitats, waste dispersal hydropower generation, tourists' activities and of course, supply of potable drinking water. However, no one can say in definite terms, the irresistibly competitive uses of this sort, instream or offstream. It is on account of contemporary concerns on water that various aspects are coming up in regard to water resources. Obviously, our query should be how these resources are related to floodplain management. Further, if water resources management is well kept up, why is then a separate concern of floodplain management ? Is it that the floodplain management provides an overarching framework ?

We have discussed a bit of floodplains in the earlier lines. Speaking from a viewpoint of totality, floodplains are those low lands linked with channels of rivers, streams, lakes, the shorelines of oceans or other bodies of standing water. We mention these as they provide the lands being frequently inundated by flood waters. Floodplains are also characterised by various processes such as climate, erosion, deposition, hydrology, extreme atmospheric and oceanic events etc. One can, indeed, identify a variety of natural and cultural resources of floodplains. We have already mentioned water resources ; others are living and cultural resources. These three resources, in particular, can be further suvdivided in details so that the dynamical systems that contribute to their support become discernible. In fact, the tremendous resource value of floodplains is often glossed over : the way these support and provide diversity of flora, fauna, fish and wild life and also maintain a harvest of natural products, contributes immensely economic aspects of benefit provided by floodplains. We have already discussed wetlands as most prominent of floodplain resources. Of late, coastal wetlands have come into the picture because of there being considered as part of estuarine systems which include tidal marshes, mangrove swamps etc. which, it is believed–could have withstood the onslaught of tsunami waves in coastal areas. Resource management in its entirety should also be brought to bear on management of all these resources. It is fairly well known that various forms of flooding can occur depending on the location, topography, soils and weather conditions. That the coastal flooding can be caused by tsunamis is well borne out by happenings in recent times. One can work out the losses on account of floodplains. Floodplain management in different countries in the world has arisen in response to prevent floodplain losses. Most of these efforts seek to control flooding through structural measures, incorporating innovative ways for analysing and predictive flooding with, of course, attending to the natural and cultural resources of floodplains. It is being increasingly felt that flood losses and disaster relief costs continue to rise because of antinatural occupancy and abuse of floodplains of the country concerned. This calls for approaches from a wider standpoint, such as zoning, landuse regulating, flood forecasting, relocation of property and alternative water storage techniques. There should be fresh efforts on flood plain information reports providing a portrayal of a community's flood hazard from a wider perspective. Why shouldn't there be comprehensive river basin planning and flood control within the context of floodplain development ? There should be regular steps to evaluate flood hazards before undertaking any centrally supported action in preventing an economic use and development of floodplains. How are we to recognise the multiple values of floodplains ? Should there be reconsideration of environmental values coupled with that multiplicity ? In other words, although, the decade of seventies, otherwise called the environmental decade, has brought about laws and regulation pertaining to the upkeep of environment, the same enterprise does not as yet seem to be visible in disaster management, particularly, those related to water. The floodplain management has to be updated from time to time so that disaster relief contingent upon mitigation is well taken care of. Managing floodplains is thus a continuing evolutionary process ; infact hazard mitigation planning has to rope in proactive engineers and supporting personnel so that participation in interagency hazard mitigation becomes worthwhile. With these observations, let us look for a conceptual framework, of course, with real applicability for floodplain management.

Floodplain management, although, conditioned by the concerns of flood loss reduction, depending on flood control, ought to stem from well planned initiative, evolving methodologies and congenial environment. Any indictment on the unwise use of floodplains will not simply enable floodplain management framework to move on its own ; the conceptual framework has to have multi-objective approach to use of floodplains without losing sight of the need for reduction of losses on natural and cultural resources that need to be identified and monitored from time to time. How to reckon with values for protection of natural and cultural resources ? How to strategise activities for implementation ? How to go about with accountability and responsibility of roles in different echelons of the policymakers, NGO's and also the private sectors ? Essentially a strategy pertaining to floodplain management has to bank on (a) altering proneness to floods and consequent damage and devastation (b) changing the nature of flooding (c) lessening the impact of flooding on the community around (d) restoring and preserving the natural and cultural resources of floodplains.

Any sustainable programme on floodplain management must reflect some guiding principles in a country like ours with state and central responsibilities. Prior to this, there should be an overview of floodplains in the context of community, region, local and national necessities. While goals of an effective floodplain management are to be identified, these need to be distinguished, strictly, from objectives which should necessarily consist of economic value addition, environmental ambience and socio-cultural sustenance. A sound principle in this regard ought to reflect futuristic needs and the role of floodplain. Developmental alternatives should be kept in view after assessment of flood losses and so, seeking alternative strategies so as not to aggravate such losses, of course, continues ; monitoring and apprising the public should be allied features. While the cost benefit analysis needs to be accounted for on the basis of impacts on floodplains, the behavioural pattern and functional styles of decision makers should be revealingly discernible so that inter-and intra-coordination among the agencies do not go by default. This is, by and large, the essence of any minimal framework on floodplain management ; of course, amenable to flexibility in the light of applicable to regions and community, as well.

Besides the above, one should mention flood proofing which is in a way the use of permanent contingent or emergency methods to either prevent flood waters from getting into buildings or to minimize the loss, if water gets in. Some of the methods are the use of water type seals, closures or barriers with water resistant materials which may require, for a short time, the relocation of the materials likely to be affected because of flood waters. There are also structures built up well above the average levels of flood waters getting into buildings. A serious component of contemporary floodplain management is to have floodplain structures so as to accommodate floodproofing and/or construction of elevated structures in accordance with appropriate rules and regulations. The issues relating to strategising structural measures for altering the flood itself are fairly well-known. There are equally a few concerns how to be vigilant about such imperatives without being too expensive about structural measures ; there seems to be escalation of investments while implementing the imperatives of flood control. There is also a fresh thinking about competing uses of the reservoirs and dams, where as levees and flood walls are also taken as improvisation. The risk element of a potential dam failure has to be assessed in the light of the severity of potential impacts and one often tends to distinguish between a high hazard and a low hazard for dam failures. If there be flooding due to storm water, flood management acquires a new dimension and one often seeks alternative approaches for the purpose; for example, run off may be taken care of within a site so that one can manage the total run off within a watershed by allowing discharges from different sectors to reach the main channel at different points of time, reducing peak flows down stream.

Although in contemporary times, particularly, in the context of recent tsunamis in the Indian shorelines, structures like seawalls, bulk heads etc. are being pleaded for protection from the onslaughts of flooding, there is scepticism about whether such measures will reinvigorate or widen the beach. One should go in for low cost and small scale measures. Vegetation plantings are rightly suggested besides the above sea barriers. Floodplain management can in no way be indifferent to land treatment measures.

We have reecognised few terms that go a long way in managing floodplains. These are essentially leading heights of this kind of management. We can no way rule out the vital component of awareness and perception about the value of natural floodplains and accompanying this, perception of flood risks. Although it may sound a bit elitist, we can hardly afford to lose sight of parameters of hydrology and geomorphology that are vital to floodplain management. Forecasting and warning using remote sensing and satellite techniques can in no way be glossed over. We haven't, as yet, a wide use of GIS and the floodplain management data base. In general, requirements of science and technology inputs in flood management are not often taken into consideration. Measures of flood management depend, in contemporary times, on utilising the space-based remotely-sensed data without being apprised with ground based data. It is a part of the reality that work on flood mapping using satellite data has an excellent and fruitful spin off. Flood control and management depend heavily on data collected by National Remote Sensing Agncy (NRSA) ; even forecasts on flood damages are based on multidate satellite data. Landsat data are certainly significant to flood processes, studies and inundation prediction model. One can multiply such attempts in different countries of the world ; for example, the flood reducing system in the West African region has revealed imagery on the METEOSAT data of European Space Agency (ESA).

There are many other emerging aspects because of the impact of information technology that need to be made use of in ways so that emergency services become aware of the risk in potential disaster. For example, in dealing with Probable Maximum Flood (PMF), hydrologists use to calculate the PMF for individual dams from regional values of the heaviest possible rain, the Probable Maximum Precipitation (PMP). Despite a variety of ways to prevent dam failure, it looks that even now appropriate use of relevant computer programmes for the data base on such assessments is far from being precise but how can we dispense with comparative magnitude that indicate real risks involved ? The community still keep on questioning whether the emergency services know the areas downstream that would be affected by the flood water. Who would warn the public ? How can people be evacuated ? Going further ,can we say that the governments should establish the degree of risk to those living downstream of a dam ? If a dam fails for some reasons or the other, can we have, well before that event, detailed maps of the zone that would be inundated by dam failure? How to obtain an estimate the speed of the flood water ? Dosen't it indicate the time between the failure and arrival of the flood waves at different locations downstream ? Isn't it the kind of analogue of the recent tsunami which did indicate a fairly long interval for emergency services to provide warning and evacuation processes ? The computer programmes that deal with the hydrology of such failures need to be updated in the light of recent methodologies. Apparently, such analyses seem to be within the domain of engineers and the technologists but the emergency services need to be totally transparent about inundation maps and even, dam safety measures. Emergency planning on this score should be improved upon not just purely on conventional structural and nonstructural measures but also on developing continuing rapport with the community so as to whether the efficacy of warnings and evacuation have impacts on them more than what might have happened earlier.

Drought Management

The Indian setting has not, by now, seen any drought disaster response to it and management, better than what happened in 1987. The details are well documented in the Government of India publication in two volumes following the drought.Despite the ways, the drought disaster problems have grappled with vexed issues and perhaps, in a way, polemics on drought are continually on; scarcity becomes automatically the point of contentious issues among a wide varieties of actors in different sectors of the society. 1987 had its immediate precursor in 1985-86. Water and food grain production come up incontestably within the ambit of drought management. The Indians seem to be fairly wedded to a worst kind of drought as the one which affects both the seasons, kharif and rabi, often coming in close succession of an earlier drought year in drought-prone and non-drought prone areas. What is relevant now, for emergency managers is to identify that drought inflicts misery on the bulk mass of the rural society for whom the pretext of climate or rainfall causing droughts hardly makes any sense and more so, in the context of 'entitlement view' expounded by Amartya Sen. One can develop

Fig. 9.2 : On Droughts (Courtesy : New Scientist)

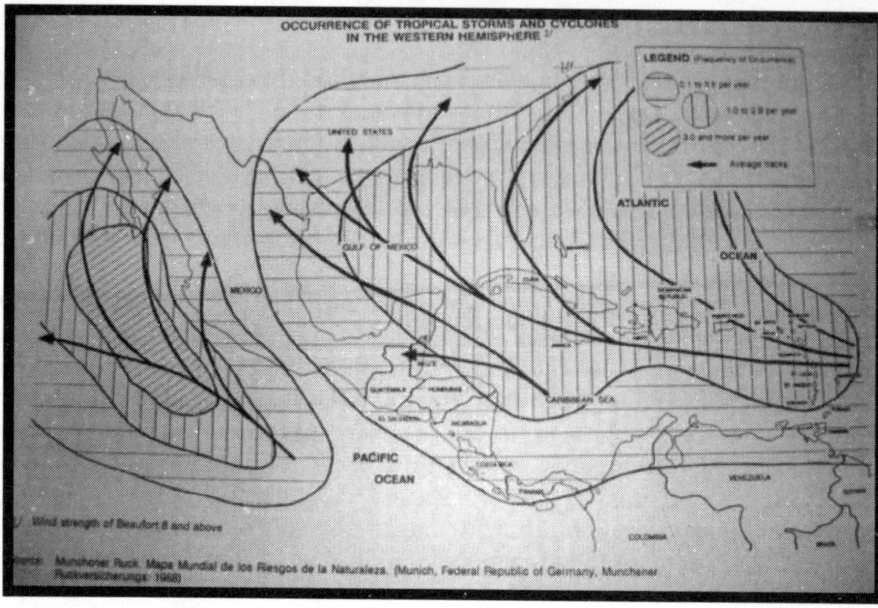

Fig. 9.3 (Courtesy : Organisation of American States, USAID, Washington D.C.)

this along with the diversity of disasters caused by droughts affecting cultivation, migration of cultivators, diversification of economic activities etc. One has to take necessarily a long term perspective drought management in the context of countries like ours. Whenever a long term analysis is to be kept in view, we cannot gloss over likely imbalances in eco-systems, here, in drought affected areas. Also, here, the crisis management does not pay dividends appreciably because this is conceptualised appropriately; quick response may lead to hastening up ill-prepared schemes. It becomes almost a convention with the establishment to go in for efficiency and effectiveness, without questioning the validity of officially time-honourd goals or objectives of the policy. There cannot be any dearth of statistics, in respect of supporting assessments. The restoration of ecological balances should certainly be one of the areas of great concern and action. NGOs are, over the decades, putting in efforts to combat harsh realities of drought affected areas. It often looks that the tenets of disaster management are sometimes not adhered to, phasewise ; recovery and relief occupy greater attention than others. There does not appear to be any enterprise on an emergency and contingency basis, as with other slow onset types of natural disasters. It is presumably because of social factors that management of droughts has to have distinctively different contours having interfaces with climate, food, agriculture, water supply agro-climate, aridity and the like that impinge on the economically weaker segments of the society. Drought management has to be well apprised of creeping character of the crisis, rather than being declared manifest overnight. The community suffering from drought afflictions cannot but be well on the guard of being infected with drought ailments. In a way, the community needs to be equipped with capability of managing themselves in collaboration with voluntary and non-governmental organisations ; of course, with a continuing and substantial support, in financial and social terms, from the establishments that deal with such calamities. Capacity building and providing restoration mechanisms should be the mainstay of ventures on drought management. The emerging issues pertaining to scientific and technological aspects of confronting with droughts should be heeded to so that risk and vulnerability are well taken care of. That practices on drought management are largely social, does not need any additional emphasis ; further, there is a societal genesis of human behaviour towards droughts so as to give rise to what is often called drought culture but this is a typical scientific or technical approach that occupies attention and so, responds on one aspect of the scenario, either physical or social. Its physical and human dimension, are intertwined so that a societal view point becomes all the more needed when we look at drought from people's perspective. We haven't mentioned about the impact of climatic change on droughts, in that their frequencies, intensities, durations and future locations are reasonably altered. In view of these arguments, we are in need of designing, precisely speaking, methodological approaches capable of managing drought situations. It looks that the most methodologically balanced exercise in this direction should have necessarily historical and structural focus. Social reality in the context of such a methodology is always apolitical but at the same time, quantitative. It is often a mundane affair to view the drought striken only as victims but, can't some one be capable of aligning himself or herself with innovations ? Perhaps, one can do so, provided his or her social identity is not in question. The governmental approach and action is, in general, a socio-political phenomenon. The establishment should necessarily be a watchdog for, in a way, "drought industry" foisted, by powerful groups, who benefit from drought by preying upon drought-striken poor and making money out of it. Analogously, the establishment should encourage the drought culture, which, in a way, is the core around which people can integrate into their framework of fundamental values through internalising the afflictions of drought-striken community so as to lead to creativity of community, affected or otherwise through adaptive norms. Therefore, an important facet of management, at the community level should be to work which the NGOs and voluntary organisations and to insist on the participation of the vulnerable community in any mode of public system on drought management.

10

EPILOGUE : ENVISIONING NATURAL DISASTER MANAGEMENT

Introduction

In the foregoing pages, some glimpses of natural disaster management are delineated along some reasonably structured directions. There are, doubtless, some areas which cannot be touched upon mainly not because of having acquired a character on maturity, some are pressingly urgent but yet to acquire an arguably cohesive structure. Therefore, there is a need to dwell on some of these aspects that would definitely call for treatment with a management flair in the decades to come. These facets require presentations with a flavour of skills for management and this is what is being precisely set forth in this chapter.

Risk and vulnerability being areas of greater concern than what they used to be earlier, they need to be considered afresh largely not because of an integrated disaster management which did not figure earlier and will be dealt with now but for the context of coastal vulnerability which Indian ocean stands terribly exposed by the tsunamis. The un-ness being a vulnerability irretrievably associateed with every natural disaster, how do we manage the unexpected, remaining as far as possible unfazed with the threatening natural disasters ? This finds consideration here. Vulnerability of the community is harped upon ; in this context, how does communitarianism play an adequate role in the management of disasters ? How does self-management explore doing so ? Can volunteerism play as a vital factor in such endeavours ? In a way, how can an emergency manager empower a community to manage themselves ? Obviously, these are posers generated by the presentations in the earlier chapters. That their answers provide a good feedback in terms of management, is sought to be developed in this chapter. A futuristic view with potential managerial ilk is also discussed. Finally, some relevant issues, often raised here and there, are addressed to, answering them to the extent they are aptly amenable. With all the above observations, one ought to confess that some well structured format should be provided or at least indicated so that realisation and/or concretisation can be a part of the realities with familiar overtones. Hence, few lines below on planning and development, so as not to have foreshocks with inadequate measures on preparedness, mitigation and response. It will equally be in the fitness of things to refer, to some issues for reasons of continuity, to some observations on crisis, and planning as described immediately below.

Revisiting Planning on Crisis Management

It often appears that planning on crisis management has been somewhat adhoc and far from being comprehensive. It is also found that one kind of disaster may have a close similarity with another which may not have happened in the same area but at a different points of time. For example, the floods in the Andhra Pradesh or Orissa or in Tamil Nadu or the floods in all the three states, form as though a vast inland sea behaving somewhat closely like a mini-tsunami. In fact, loss of life, extensive damage to property and livestock, loss of money, often an enormous loss of crops all, as a whole, amounts to few crores. Operationally one does a lot so that the disaster mitigation becomes effective but several questions may often arise particularly whether the disaster could have been avoided or anything can be done to preempt a future calamity. It looks that the process of recovery is executed mostly by engineers, overseers and the like. True, that technicians may do but because of governmental intervention the expertise of a technological kind may perhaps be of no substantial consequence, as they are often allegedly to be so. There are, of course, several layers of administration so as to allow the flow of financial support but what is of paramount significance is crisis related policy making and operational planning. As seen recently, how one can make use of defence services in the times of tsunami crisis, particularly when they are called to help shift, transport, storage and medical supplies. It has to be acknowledged that satellite communication has to be a new dimension to crisis planning so that the dependence on conventional life-line systems, vulnerable as they are, can be bypassed. The constituents of an effective crisis planning should be civil-military relationships, organisational capability and containing inter–and intra–conflicting issues and problems. A quasi-defence and centralised mechanism–has to be resorted to crisis planning and management. In order that such a planning be comprehensive, we need to look around cross-national disaster experiences and competencies; in particular; some indicators have already been mentioned earlier chapters, for example, disaster planning being conceived as a dominantly civilian endeavour. The disaster response should manifest a greater sensitivity to community and human services without being stuck up by a kind of disaster planning focussing simply around recovery. In sum, one can say that, by comprehensive planning for contending with crisis calls for increasingly effective coupling in civil-defence capacity.

Whatever be the field of activity, any kind of planning, partial or comprehensive, requires the assistance of few key participants involving planners and victims, as well. We have already spoken about contingency based modelling. This is related to decision-making processes which, in turn, may coincide with decision-making models. While embarking on decisions, there can be various alternative decision-paths ; these are likely to stem from forms of crises which, in turn, keep on creeping. There may be caveats, often analytical in nature and these may become somehow amenable to analytical treatment ; one can convert a statistical framework into one of dynamical decision-paths and this what a contingency model has as its one of the components. An important advantage of a careful study of creeping crisis is that a learning curve for one kind of crisis will provide a potential learning edge. One can extend the benefits on creeping crisis to rural scenarios, pertaining to planning and management. Rural disaster management involves issues of floods, droughts etc. There are few countries particularly the former Soviet Union or the present Russian Federation, who are involved in developing designs of tsunamis. This requires a sort of modelling which makes use of improved technology and their adaptations, too.

Complex Emergencies and Natural Disaster-Management

We know what we mean by emergencies, in close alliance with natural disasters. A close mutuality between them, by all shades of opinion, is that lives are fundamentally disrupted and often, threatened by other repercussions. Although there is an incorrigible element of history that people are uprooted by natural disasters, but man has never, at the same time, been so strong and so vulnerable in the face of such disasters. Many of such eventual scourges afflicting mankind are of such a deadly nature that they can be labeled as complex emergencies. Obviously, they can emerge from a variety of causes. The percentage in the number of complex emergencies keeps on escalating but it is not that the percentage increase is alarming. Rather it is the disaster-driven emergencies that are most revealing. More and more, we see around the emergence of complex emergencies stemming from some kind of terrible breakdown of structures, communication systems, besides deprivation of livelihood and means to survive. People, often, seem to be trapped between conflicting realities : they are thrown out of their homelands and they are to flee somewhere for their survival. Haven't we heard in the recent tsunami, a sort of upsurge of ethnic practices among the indigenous populations in the Andaman-Nicobar islands ? True, individuals, organisations and agencies imbued with communitarianism, have to spend, more than ever before, time and energy to provide humanitarian relief and aid to the helpless victims of complex emergencies. It is equally true that whatever be the nature of natural hazards, they cut across our economic, social, political and cultural divides in our society, so that the totality of response is largely shaped by the extent to which proactive programmes can be effectively concretised. The latest tsunami shows, fairly in abundance, the systemic nature of natural disasters and revealingly, paucity of managerial skills, particularly, in respect of information exchange but also an increasingly sensitive understanding of the possibility for greater regional or global collective action on matters, definitely, of global concern. Let us now seek the conceptualisation of complexities in this direction.

One often makes a conclusion that passing on data inputs and outputs is an essential element of disaster information management : it is far from being a linear process but an interactive process of knowledge building on the basis of information exchange within a very complex system of social, institutional and infrastructural networks, which seem to be disturbingly lacking in the case of recent tsunamis. Disaster management, as it stands now, seems to be often, not having fully exhaustive attributes for making it fairly definable, although the complexity is well articulated by the causes, probable risks, frequencies and consequencies of disaster. One cannot do away with cumulative knowledge of familiar natural disasters such as floods, earthquakes etc. It often becomes a pity that institutional responses are reactive rather than proactive which unfortunately comes in the way of recognising the need for an integrated approach. Even within response planning, there often appears mis-conception about the delineation between an emergency and a disaster. The community seems to be suffering because of the lack of an integrated approach, as it falls upon its resilient capabilities rather than depending on the support from mandated service organisations. The complexity can well arise from institutional barriers to information access or disclosure. It is now well accepted that the most effective way to disseminate disaster-related information is by responding to the real rather than to user's sense of receptivity. Information management process, an integral part of disaster management is, thus, a dynamic interactive process where interdependency is an enabling element to construct a basic framework for an integrated disaster management. Interdependency does, of course, exist in one way or the other and the objective will be to allow it to have exchanges well beyond currently accepted boundaries. Once complexity in the arena of natural disasters is reckoned, one must necessarily take steps in learning from complexity consisting of understanding it, assessing it and managing it, at all levels. Obviously, these are terms endowed with genericity. Usual laws of casuality do not, obviously, hold here ; it is not because that casuality has ceased to exist but we are in a world that is driven by connectivity and universal access. Hence, we cannot shy away from global cencerns. In a disaster, we have, tremendous pace of change which is not confined to the conventional commercial processes of production and distribution. Here, the takers become, in some ways, beneficiaries of some sort of support without being given an opportunity to be choosy. In the few lines above, we have talked about the value of information for the affected community ; we know it for certain that the value of information has an exponential growth when individuals become a part of the interactive community and this is also the case in disaster enterprises. Speed, intangibles and connectivity dominate the disaster landscape ; how can we devise some forms of metrics that evaluate performance in matters concerned with disasters ? There is no process as yet by the takers so as to have value adding experience in the context of natural disasters. But in the context of WCDR, we need to create a capability to operate effectively and adaptably in the changed setting of disasters. It is really managing human, physical and intangible resources so that a set of rules is learnt. In the pursuits of natural disasters, there have been few paradigm shifts in the way in which human beings make a living in the face of threatening reality. Now a days, we have a number of knowledge breakthroughs. Global impacts can hardly be bypassed while knowledge breakthroughs co-evolve to create new forms of knowledge and so, too, are organisations. Achieving success in disaster endeavours with such complexities is pretty onerous. There is an in-built nonlinearity which affects all types of involved networks. Complex evolving systems, in general, share some commonalities. We have mentioned some of these common characteristics, such as, nonlinearity and adaptibility earlier and a bit about of the characteristics of memory which relate to history of disasters in the relevant site.

Resilience of the community is fast coming up, even in the context of a natural disaster which is an analogue of "self-organisation", an important characteristic. Creativity goes along with complexity in corporate affairs but hasn't come up, formally ; but one may look for it from individuals or groups of individuals in affected or unaffected communities. The success stories are certainly related to ingenuinity of efforts which can hardly be cut-and-dried techniques. Whether the activities in different phases of the disasters have value adding characteristics, is yet to be taken up seriously. A community or an organisation continually learns how to satisfy the changing necessities of takers and givers. Success on such scores will be achieved if core competencies to deal with the future disasters are developed, inclusive, of course, of the processes utilising knowledge and resources. The corresponding vision-scenarios have to be brought to the fore. In brief, one can say that we have to seek the essence of complexity science so that strategic management practices keep on changing radically so as to cope with the current pace of change. The anticipatory phase of any natural disaster, flexibility and knowledge resources ought to be the principal characteristics of an organisation capable of contending with the disaster world, beset with complexities of natural disasters. Really speaking, pre-disaster phase is the primary key so as to provide the lead time required to create a bright future for the affected group. What is required is the requisite variety of options centered around core themes on disasters, which can, in turn, be made use of under possible future realities. The complexity perspective on natural disaster management should necessarily be a function of both, who we are and what we are capable of envisioning. The antenna of complexities provides an extraordinarily novel way of thinking about disaster phenomena which is essentially an evolutinary thinking ; such a type of thinking brings out from the past history, some unpredictable and often improvable truth surrounding us in the garb of failures. There remains then the consideration of the complex nature of leadership versus uncertainty. This matrix is often taken as a 2 × 2 matrix with the community task, leadership task, operational task and technical task as elements of this matrix.

Coastal Vulnerability Management

Coastal resource management and coastal development continue to be areas of great concern and interest. The rationale for such endeavours is in no way far to seek. Like other coasts, the Indian coast is indescribably replete with wealth of valuable natural and cultural resources. Obviously, there has been no dearth of coastal harbours, historic and archeologic sites, beaches, salt marshes etc. These resources provide much-talked about wild life habitat, magnificient scenic beauty and open space. Now a days coastal scenarios provide stimuli for generating financial resources, dominantly through tourism industry. Apart from these, coastal resources provide protection from wave action and serve as critical buffer structures from coastal storms, tsunamis etc. That is all the more why such resources and zones are required to be protected and monitored. Coastal vulnerability has once again been raked up by the recent tsunami. Vulnerability finds a greater prominence than what it used to be earlier, as evidenced by the deliberations of WCDR at Kobe in January 2005, fairly closely on the heels of the tsunami disaster. The Hyogo declaration has put equal stress on development. Vulnerability reduction apparently becomes a matter of urgency. This poses the question: how do vulnerability and development go together ? Development as a term has acquired a greater ambit than when it used to be referred to as connoting "betterment"; over time, the society seeks to sustain growth. Vulnerability is seemingly a reflection of dependency that runs contrary to "betterment". Does the difference between betterment and vulnerability reflect the resilience of the society to natural disasters ? Perhaps in some ways it is, but when a given event makes the stage of "betterment" to fall well below the stage of vulnerability; some extra inputs are warranted so that the path for "betterment" may suffer for some time to come. The natural disaster can worsen its fragility not only in terms of density of populations and their wherewithal but also their dependency on others so that the loss of development gains is incurred, thereby incapacitating the society for sustaining the avowed increase in "betterment". Coming back to coastal resources, we need to look for indepth analysis whether we can afford to have sufficient resources to restructure the momentum of development in the post-disaster phase. Vulnerability and more so, coastal vulnerability, provide opportunities to look at the coastal populations. Modern developments in science and technology have enabled us to identify goods, services and hazards of eco-systems which are terribly prone to vulnerabilities. Vulnerability reduction is thus an important measure for whom development is being aimed at. Coastal vulnerability has, of course, many commonalities with seismic vulnerability or flood vulnerability. Some thoughts and ideas on coastal vulnerability to tropical cyclones (TC) have already gone into extensive oceanographical and meteorological knowledge on air-sea interactions, storm surges and tides and on modern forecasting techniques for extreme TC winds and exceptional TC precipitation, alongside materials on the engineering of disaster proof structures etc. Sir James Lighthill, as Chairman ICSU-IDNDR, made a strong plea on pursuits on coastal vulnerability in regard to TCs.

The recent tsunami has not merely brought up the issues of vulnerability management but additional problems with innovative characteristics of the coupled tsunamicity. Some organisational procedures have gone into combating tsunami disasters but obviously in conventional ways. One has to think afresh whether such organisational models, dominantly top-down in character, will at all work for the future. Local communities or indigenous communities in the islands have responded, in recent times, to do something on their own and that, too, with some amount of success. The governmental and non-governmental supports, often reach them well behind what they should have, atleast in securing shelter and rescue. In the terminology of the organisation, public or private, of course, in the national arena, their responses go by default. Honestly speaking, organisational behaviour is often at stake, mainly because of a host of factors such as coordination, connectivity and cohesion in respect of responsibility and accountability. Organisational behaviour, by itself, is a vast field, involving not only the individual and collective human behaviour but also the environment around the organisation. The point at issue is, now, to come to terms with abruptness of changes in organisations, in the aftermath of a natural disaster. In managing such changes, it is necessary to identify the focus of change. But how do organisations know whom they should change ? What sequels should an organisation look for ? There are not immediately available answers to these questions; yet there are signals that the changes will take place. Awareness of factors enforcing changes should be continually looked into. One could always develop a generic typology of organisational change in the context of natural disasters. Concept, complexity and cost-appear along with the possibility for resistance to change and uncertainty, as well. There lies the need for planned changes. There is, of course, transition management which is the totality of processes of identifying planning, organising and implementing whenever the change occurs. It is not that an emergency management is a transition management but has to take on sometimes, the latter role ; this may include, if so necessary, an interim management for ensuing continuity and control during the transition. These subtleties are being specifically pointed out as they ought to occur in the changing formats of natural disasters. Isn't it that the forecasts and warnings involve changes to be communicated to the community around and so, communication in such situational junctures has to play a key role in transitional management ? While one can identify a host of processes, in respect of management changes, a holistic view can hardly be dispensed with, for that keeps disaster management in operation. In other words, management strategies may often have to be structured so as to meet impediments to required changes. One can suggest a number of organisational models for disaster management. The recent attempts by the fishermen and the closely allied communities, have shown their experiences to such a great extent that one can adopt a bottom-up approach to emergency management and disaster relief so that a local grassroot level segment of the society becomes primarily responsible for attending to preparedness and mitigation needs. This meets vulnerability reduction in a period when transition management can suffice. As shown by recent events and actions, too, families and whole communities may have access to less vulnerable areas but these need to be strengthened by a larger segment of a society with government sponsorship. In matters of coastal vulnerability, being of a wider and more sensitive type, each phase of the disaster management cycle, should address vulnerability and vulnerability reduction issues till the real figures and financial terms can accomodate a special vulnerability level and corresponding mitigation measures. This becomes aptly relevant to the dire necessity of installation and upkeep of a warning system in the Indian ocean. It is not the Hyogo declaration of 2005 at Kobe that gave indicators but the recent Afro-Asian Summit at Indonesia also emphasised on the same. A close cooperation between the corporate and the government sectors, may be aimed at.

Any discussion on management of coastal vulnerability remains incomplete if the pressing concern on coastal zones goes by default. This is all the more important for Indian Coastal Regulation Zones (CRZ) guidelines which came up under the Environment Protection Act. It is not the Indian concern alone ; indeed, some other parts of South Asian shores have similar stakes ; for example, the shores of Maldives, the coastlines of East China Sea. Till the devastating exposure of the shores in the wake of the tsunami disaster, there seemed to be a kind of complacency about the management of coastal zones. It is quite natural that second thoughts are, of late, arising in regard to the adequacy or otherwise, of the upkeep of these coastal zones. There is a genuine criticism about lack of early warning systems that, had, if existed, could have effectively alerted coastal communities and establishments, as well. The last Afro-Asian meet in Indonesia has confirmed, in fairly categorical terms, to the intention to have an effective warning system in the Indian ocean, on the basis of what may be called an Pan Asian Framework of scientific collaboration and that's definitely for impending disasters in future.

It is being increasingly felt that India's management of coastal regulations haven't stood the test of time, as evidenced by the high-scale damage induced by the recent tsunami. The general trend of allegations seems to lie in providing permissions given to land reclamation, aqua-culture, hotels and tourism projects etc. in coastal areas. Environmetalists cannot but be sore about them, particularly because of the severe onslaught on mangrove ecosystems along India's soft coastal areas. It shouldn't be taken to mean that the mangroves are not always being conserved ; indeed, coastal lands are often remarkably resilient to natural calamities. What has then come in the way of firm implementation of CRZ guidelines ? Have they happened only in Indian shores ? What is the global trend on management of coastal zones in this regard ? Why can't we rope in the fishermen community for their protection, in the wider context of maintenance of their rights and livelihood in coastal areas ? These are aspects of in-depth thinking in this direction, particularly in the light of developing, what is often called "a natural resource continuum". The classical views and values on conservationism continue relentlessly. Coastal zones stand exposed to vulnerabilities, may be worse than that have happened in regard to tsunamis. Therefore, development becomes a compelling necessity and the general propensity of this votary of thoughts on development is to keep on supporting economic development in coastal areas through tourism, commercial agriculture, fisheries and a variety of industries. Maldives has gone much ahead in this direction. China seems to be creeping for economic viability of coastlines along East China Sea, but fortunately Indian CRZ guidelines have not so far allowed to be victims in this direction. But, isn't it that there are some constructional and so, structural disabilities, somewhere in CRZ guidelines ? Any confusion, on that score, will lead to terrible competitiveness among the market forces, which will distressingly will leave no stone unturned for exploitation of the traditional communities on the seashore. Why can't we empower these communities in the management of natural resources in shore islands ? There has to be somewhere a mission mode in regard to coastal management. There are few economic implications of assessing coastal projects ; as is often said, we are yet to have assessment of projects ; involving coastal areas on the basis of "critical value judgements". It is often said that fishermen communities in North Kerala suffer from seawalls constructed in areas. So, in brief, management of coastal vulnerabilities has to be immune from infirmities, particularly in supporting unsustainable tourism and real estate development in coastal areas. Hopefully, the natural disaster policy being envisaged through a central intervention, will come to terms to these weaklings of coastal areas.

Facing the Un-ness of Disasters

The previous chapters have given broad-overviews of what disasters are all about and what implications and repercussions of their management may have. As per the official coinage "un-ness" by Hewitt [11], one should dwell on its management from its variation in multiple forms. There is every likelihood, as is often said, that one takes recourse to mostly skills somewhat characterised by naivety. The spur of time or the instantaneous response may result in previous state of preparedness. If a crisis is to pre-empted, one may resort to such practices without being equipped in any way earlier. But the variation of "un-ness" in regard to a disaster, seldom occurs separately. If these were the courses of happennings continuously or the other way round, one can confront each of them in ways and means that may sound fairly well familiar. The ground reality is the totality of "un-ness," often without being completely exhaustive. If the "un-ness" of occasions could be split into self-contained units, there can be the use of the well known principle of reductionism. Therefore, strategisation of management for a disaster, in terms of variation of "un-ness", calls for a different approach. A juxtaposition of encountering the variation of "un-ness" may suffice for a transitional management, but in no way, any format of an integrated management.

As the artifices of reductionism will not arguably work here, one should look for some kind of structuralization in the totality of disaster management. The point at issue, then, becomes how to structuralise disaster management. The initial step, obviously, should be to conceptualise a structuralized approach to disaster management. We may draw upon a variety of methods used for structralisation in other disciplines. But before we do so, we must ensure that there is not a shred of arbitrariness in the system, for the ultimate objective will be to provide a perspective on the management of disasters, as governing modes of sustainability. Natural disaster management must necessarily include disaster assessment comprising all phases of the disaster cycle. A survey of figures of losses, damages etc. which, of course, keep on escalating, will not bring about the conceptualisation of the core sector. One can hardly deny, that some disaster events have happened so surprisingly that these have led to changes in ways in which the so-called previous ones could be managed and so, flexibility in approach to emergency management system is a must. Any quick glance on the findings from the classic literature will reinforce the statement that, managerially speaking, there is an unique flair. Even though the establishment and the community may take up measures, for example, developing new technologies for protection, shelter and rescue, each of these segments should see to it that such measures do not become a source of added vulnerability in future. But disaster mitigation is the significant phase in which upshots should become the cynosure of what can possibly happen as extreme future events. Emergency preparedness, ideally speaking, comprises activities to be undertaken before a disaster impacts so that there are enabling features for the social organisations to respond through activity when a disaster strikes and this requires, obviously, organisational preparedness activities. Emergency response, however, becomes an ability to undertake action for a short period, prior to–during–after-disaster impacts so that there is considerable reduction in disruptions and losses. We have said earlier on many aspects of post-disaster three 'r' phases. The recent tsunami can be said to have a spatial as well as a temporal dimension ; for, there are regions, that witnessed relatively localised areas of severe damages and devastations, as in Sumatra and areas where impacts were spread over a wide geographic area, such as Sri Lanka, Andaman & Nicobar islands, Maldives and Indian shores. We are yet to have a report that will certainly show to the public that tsunamis did produce a range of impacts that can be described as direct, secondary or disaster-induced and indirect effects. We would soon hopefully have the geography of the disaster-event as involving a series of concentric zones, where the common centre gives the area of very severe impact surrounded by fringe areas, as in Indian ocean in which there were also alarming losses and damages. One can multiply this range of negative impacts from various points view, not exclusively business, social and cultural. One can afford to conceptualise disaster response intended mainly for organisations, whose capabilities are often contested, because of there being an escalation of demand. Isn't it that's an excessive demand, unexpected? Can one argue that unanticipated demands exceed the capability of organisations to respond? A routinised response will not suffice and improvisation has to be resorted to, for coping measures. A structuralised view should put accent more on emergency preparedness and response because natural disasters vary in frequency and temporal regularity in regard to rapid onset disasters (for, slow onset disasters are accompanied by environmental degradation). As in tsunamis, a larger warning period enables organisations to issue warnings to the public. This, obviously, increases the response capability which, in turn, tells upon poor emergency preparedness. Response and forewarning are, thus, inextricably linked and every response activity can be effective and losses to be minimal, if the forewarning becomes a continuing reality. One can end up by stating that physical features should be considered as important defining characteristics of conceptualisation of disasters ; there can always be difference, if not dilemmas, importantly significant for understanding of emergency preparedness and response activities. These remarks do not take a lighter view of practitioner's conceptualisation of natural disasters, even though they may often be irksome.

A major strategy of doing away with "un-ness" should be to encourage people to prepare, that is, in a way imbuing people, in families with emergency preparedness. Ethnic groups are fairly equipped with this ability as is shown recently by the island community. What may come in the way to put into use hazard-related knowledge is their nomadic character.

Since we have identified emergency preparedness as a main component for structurisation (so as to do away with "un-ness"), a question arises about the un-ness of preparedness so as to be capable to respond to disaster situations. May be data, corresponding to specificity of disaster events are not that available but, pre-planning in the household sector, seems to call for adaptivity. Quarantelli [23] remarked, "even if preparedness is good ; it does not follow that managing a disaster will also be good ..... Good planning does not automatically translate into good managing". The question still remains whether the effective response takes place because of good planning, emergency planning or sheer coincidence of circumstances. Having felt such a sceptism, in order to dispense with "un-ness", it seems desirable to look for optional ways to organise disaster response. We can well recall what we mentioned in earlier chapters, the relevant remarks about emergency measures with success. Whatever be the adversities of planning one may talk about, there is a greater need now, than ever before, on delineation on general principles of planning, applicable to a wide range of planning efforts of natural disasters. If disaster management is to have a continuing existence, free from adhocism, we need to situate hazard and disaster management policies in their social contexts, so that a better understanding of activities, programmes, and indepth pursuits can become generalisable from one event to another, of course, with event-related specificities. The totality of events and in the Indian context in particular, like seismicities in Latur and Bhuj and tsunamicity across the Indian ocean, demand a strictly macro-social view shaping the disaster management strategies of the governmental agencies and the society at large. Professionalising emergency management, in terms of choice, particularly structuring them in the hazard domain, can well facilitate managing hazards and their impacts on the community; the community can only then be kept fairly immune from fragilities of all kinds that may or may not concur with disaster events.

Volunteering Counter Disaster Efforts

In any emergency response period, one finds, frequently the work of a fairly large number of volunteers in a variety of activities related to grappling with problems of natural disasters. There are many case studies related to disaster volunteerism. But most of such participations do not often reflect strands that are being assumed fast in the area of volunteerism. It comes within the purview of management pertaining to natural disasters, as volunteerism has come to be, of late, as a model of the decision to volunteer in such enterprises. Volunteerism is, by all shades of opinion, has come to be recognised as a pro-social behaviour which is described and defined as the behaviour to extend some benefits to individuals or a groups of people. For natural disaster management, this behaviour has to have characteristics of volunteerism for a definite period, may be often for longer period. Volunteerism is a planned action and people think deeply about it before they plunge themselves to volunteering and so, this is a sort of intervention that can always stand by in emergencies. It has to work within an organisational framework. Volunteerism is to be looked upon more than a mundane affair but probably a important pro-social action rather than extending help somewhat casually to a stranger in an emergency. One can certainly conceptualise the decision of a volunteer. It might happen that volunteer-activators may labour under some sort of doubt and scepticism and instances exist in other situations where volunteers suffer from being attacked. It is true that historical events like natural disasters can serve a person's willingness to volunteer. There exist some excellent phenomena where rescuers tend to be pro-social in outlook and approach so that volunteers act proactively and pro-socially. It has become necessary, in the context of what happened about tsunami warnings that volunteers need to be roped in for purposes of dissemination of information about natural disasters. It looks that crisis-related organisations seem to be on the increase and opportunities have come as to whether we should feel the need of a volunteer for thoughtful and planned activities. There are trends in volunteerism as the world changes and, therefore, the need to maintain a strong volunteer base. Volunteers can provide the missing component of having volunteered resources on the web and the volunteers are open to virtual volunteering online services. These need to be taken into account gradually if we are to improve upon emergency planning, response and management.

Pre-viewing tasks ahead

Any sort of envisioning disaster management with all its ancillaries, namely, crises, contingencies etc. can hardly be complete if there be no efforts for institutionalisation and realisation for rescue mobilisation. Whatever the government does for a country like ours or is doing in a developed country, now a days, with dominant global and corporate participation, one should think in terms of a corporate network for meeting the challenge of disaster responses. South-Asian countries like Phillipines, which are in no way free from scourges of natural disasters, have frequently benefitted from such networks. But scepticism about the impact of the help and support often seems to be on the air ; for example, there are the overriding questions of ethics and morality. One cannot rule out the possibility of differential treatment of business group's own employees and the communities suffering as victims of disasters in the neighbourhood. This calls wholly for distilling the essence of 'corporate governance' into the functioning of the 'corporate networks' in the aftermath of natural disasters. It cannot be done simply through ritualistic joint sector move. One finds now democratization of organisational change and strategy in public and private sector organisations. Disaster management, to reiterate, is a social process by which leaders and people affected become personally implicated in the performance of their own domain of work, in the context of contributing to wider organisational changes strategies, operational improvement or even, day-to-day performances – each of which ought to be mirrored through every discernible phase of natural disaster management. Of course, crafting IT (Information Technology) has emerged as a great facilitator in such enterprises Market forces will come up in no way, surreptiously, as an effectve factor but with a changed behavioural complexion, certainly with an improved corporate ethics. Corporate governance does provide a framework as a driver on contingent scenarios, which accompany natural disasters, as described earlier. The concept of 'alliances', being confined to political or social processes, has to be drawn upon so as to acquire evolving characteristics in the arena of natural disaster management, tempered with aspirations, ethos

and realities of people in general. The Hyogo declaration's accent on risk, vulnerability, poverty reduction and sustainable development, all in the context of tsunami 2005 phenomenon in the Indian ocean, can hardly become true if interactions and ventures of the kind described above are not efficaciously strategised, whatever be the level–local, national, regional or global.

REFERENCES

(For Chapters)

1. Sinha, D. K. (Ed) : Natural Disasters Reduction for Nineties Perspectives, Aspects and Strategies, International Journal Services, Calcutta (1990).
2. Sinha, D. K. : Some Aspects of Natural Disaster Management, IRUDENDUM (2002), Kolkata.
3. Sinha D. K. and Mukherjee S. P. (Eds) : Natural Disaster Management : The West Bengal Scenario, Indian Science Congress Association (1994) Kolkata.
4. Drabeck, T. E. : Human System Responses to Disasters. An Inventory of Sociological Findings, (1986) Springer Verlag, New York.
5. Visva-Bharati : Natural Disasters : Some Issues and Concerns, Natural Disasters Management Cell (1999), Santiniketan.
6. Davis Ian : Shelter after Disaster (1978). Oxford Polytechnic Press.
7. Fisher III H. W. : Response to Disaster (1994), University Press of America, Mary land.
8. PAHO (Pan American Health Organisation) : Natural Disaster, Protecting the Public Health, WHO Publication 575 (2002), Washington.
9. Sinha D. K. : Whither Contours of Natural Disaster Management ? Asian Studies, Special Voiume on Disaster Management No.2 (2002) pp- 2-6

10. WCDR Secretariat UN/ISDR : World Conference on Disaster Reduction, (2002) Kobe, Hyogo, Japan.

11. Hewitt K (Ed) : Interpretation of Calamity. From the viewpoint of Human Ecology, (1983), Allen and Unwin, London.

12. Rosenthal, Uriel, Charles, T. Maichael (Eds) : Coping with crises. The Management of Disasters, Riots and Terrorism, p. 416.

13. Ibid : p. 437

14. Kouzmin, A. and Jarman A. M. G. : Crisis Decision Making : Towards a Con tingent Decision Path Perspective in Rosenthal, U. Charles, M. T. and Hart Paul't (Eds) Coping with Crisis. Towards a Contingent Decision Path Perspective, (1989), Charles Thomas, Springfield, pp. 379-435.

15. Lindell M. K. and A Prater, C. S. : Household adoption of seismic hazard adjustment: Comparison of residents in two states, International Journal of Mass Emergencies and Disasters (2202) 18 : pp. 317-338.

16. Wildavsky Aaron and Dake Karl. : Theorising Risk Perception : Who Fears What and Why ?, (1999), Environmental Risks and Hazards. Susan L. Cutter (Ed), Prentice-Hall of India. New Delhi pp.166-177.

17. Anderson, E. : The Risk Assessment Process in Carcinogenic Risk Assessment (1988), C. Travis (Ed).

18. Timmerman, P. : Vulnerability, Resilence and the Collapse of Society, Environmental Manograph No. 1, Institute of Environmental Studies, (1981), University of Toronto.

19. Liverman, Diana M. : Vulnerability to Global Environmental Change, Environmental Risks and Hazards (1999), Susan L. Cutter (Ed) Prentice Hall of India.

20. Sanderson David, David Jan, Twigg, John and Cowden Belinda : Disaster Mitigation, Preparedness and Response : An Audit of U.K. Assets (1996), Oxford Centre for Disaster Studies/IT Publications, ISBN.

21. Kalam, A. P. J. Abdul : Dimensions of Knowledge Society, India International Centre Quarterly (2002), Vol 29, No. 2, pp. 39-49.

22. Toffler, Alvin : Future Shock (1970), Bantam Books, New York.

23. Quarentelli E. L. : Converting Disaster Scholarship into Effective Planning and Managing : Possibilities and limitations, (1993), International Journal of Mass Emergencies and Disasters II, pp-15-39.

24. Koyzumi A. and Scott N. (Eds) : Dyanamics in Australian Public Management : Selected Essays (1990), McMillan, Melbourne.

25. Comfort L. K. (Ed) : Managing Disaster : Strategies and Policy Perspectives, (1988), Duke University of Press, Durham.

26. Dynes, R. R. : Organised Behaviour in Disasters (1970), Heath, Lexington.

27. Rosenthal, Uriel, Hart Paul 't and Kouzimin Alexander : The Bureau- Politics of Crisis Management, (1990) Vol-69, No. 2, Basil Blackwell, The Royal Institute of Public Administration.

References

(For further reading)

1. Bolin, Robert (Ed) : The Loma Earthquake Studies of Short Term Impacts. Program on Emvironment and Behaviour Manograph#50, Institute of Behavioural Sciences, University of Colorado, 1990.

2. Natural Disasters Organisation : Community Emergency Planning Guide, 2nd Edition, Australian Emergency Mannual, 1991.

3. Gupta, Harsh K. : Disaster Management. Universities Press (India) Private Limited, 2003.

4. Sharma, Vinod K. : Disaster Management. Indian Institute of Public Administration, New Delhi, 1994.

5. Sinha, D. K. : Coping with Natural Disasters : An Integrated Approach, A Task Force

Report, 18th Indian Science Congress, Indore, 1991.

6. Natural Disasters Organisation Deptt. of Defence : Report of Proceedings of a Research Work shop on Human Behaviour In Disaster In Australia, Australian Counter Disaster College, Mount Macedon, 1985.

7. Magalhaes, Antonio, R. (Ed) and Glantz, Michael, H. : Socio-economic Impacts of Climate Varia tions and Policy Responses in Brazil. United Nations Environment Programme (UNEP), CEARA, Esquel Brasil Foundation, 1992.

8. United Nations Centre for Regional Development (UNCRD) Nagoya, Japan, Centre for Integrated Rural Development for Asia and the Pacific (CIRDAP), Dhaka. : Disaster Management for Regional Devel opment Planning with People's Participa-tion, Vols I & II, 1990.

9. Pant J. C. : Disaster Management, Towards a National response System, University of Lucknow, 2001.

10. Maskrey, Andrew. : Disaster Mitigation–A Community Based Approach, OXFAM, 1989.

11. Leet, L. Don : Causes of Catastrophe, Whittlesey House, Magraw-Hill Book Company, 1948.

12. Tierney, Kathleen J., : Facing the Unexpected, Disaster Preparedness and Lindell, Michael K, and Response in the United States. Joseph Henry Press, Perry, Ronald W. Washinton D. C. American Resource Centre, 2001.

13. Jonathan Walter : World Disasters Report: Focus on recovery International Federation of Red Cross and Red Crescent Societies, Switzerland, 2001.

14. Max H. Bazerman & Michael B. Watkins, : Predictable Surprises, The Disasters you should have seen coming and how to prevent them, Harvard Business School Press, 2004.

INDEX

P

R

S

T

U

V

W